Learning Based Robot Navigation By Path Planning

BY

Tema Dino

Contents

Chapter 1

Introduction

Robotics is a multi-disciplinary area where extensive research has been done for decades. Mobile robots are one of the prominent applications of robotics. Apart from its usage on industry floors, it is used in domestic service, missions to other planets, remote surveillance, dangerous radioactive sites, etc. Mobile robots are classified based on their form factor, such as wheeled, legged, flying, and so on (Rubio et al., 2019). Wheeled mobile robots, in particular, can be effective and simple to use in flat indoor terrain (Silva Ortigoza et al., 2012). Mobile service robots have become popular in consumer electronics for the last decade. Service robots have become a part of our society, helping humans in various works such as in museums as guides, in the hospitality industry, lawn mowers, and assisting senior citizens and the physically challenged. The major reason for this popularity is that they operate autonomously and help users with daily mundane tasks.

The navigation system is an inherent part of any kind (wheeled, legged, airborne, etc.) of a mobile service robot. A navigation system is a key element in making a service robot autonomous. It helps the robot to navigate from a given source to a destination autonomously. A navigation system perceives the environment and enables the robot to navigate autonomously to perform tasks (Grollman and Billard, 2012; Loganathan and Ahmad, 2023). A navigation system comprises subsystems that perform localization, path planning, obstacle detection, and obstacle avoidance. Figure 1.1 depicts a general mobile robot navigation system. Localization is used to locate the robot's current position in the environment. A path-planning system uses an environment map or sensor to plan the robot's path. The shortest and safest global path is generated depending on the source and destination input. Obstacle avoidance identifies and avoids any static or dynamic obstacle in the path by generating a local path. Path planning is essential for an autonomous mobile robot navigation system to plan an efficient path given the source and destination.

The area of socially aware navigation has been extensively researched in the past decade, focusing mainly on planning based on human behavior (Mavrogiannis et al., 2023). This enables robots to

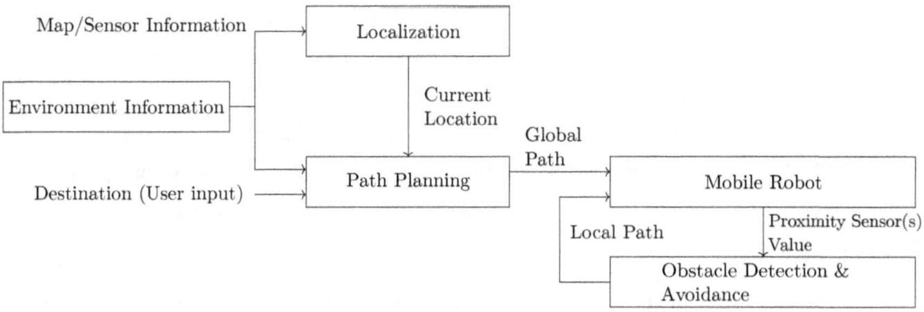

Figure 1.1: General navigation system

navigate based on human behaviors and cues. The service environment is dynamic, with humans as moving obstacles; the navigation system can predict human motion and plan the path accordingly (Kivrak et al., 2021; Bellotto and Hu, 2010; Bera et al., 2017; Brito et al., 2020). This is called local planning. While this is essential for social navigation when executing a path, the primary concern regarding environmental awareness is still an open problem (Chikhalikar et al., 2023; Hernandez et al., 2023). This is called global planning, which generates the initial path. To address this concern in this work, initial environmental awareness is incorporated into the robot using human help to train the robot to reach different destinations in the environment. There has been work in the literature that uses human help if a robot encounters problems like being physically stuck, falling, and getting delocalized are proven to be helpful (Rosenthal et al., 2011; Nanavati et al., 2021).

This work aims to develop a learning-based navigation framework for mobile service robots. A partial understanding of the environment is learned by the robot using Learning from Demonstration (LfD), through which the robot is taught to navigate one path for each destination sequentially. After learning the required paths through LfD, the robot has to generate all the possible logical paths among the known destinations and navigate autonomously. A path planner is developed for generating unknown logical paths between known destinations. This work also includes developing dynamic obstacle avoidance algorithms for complex scenarios essential for a mobile service robot navigation system. The proposed work also aims to develop a reinforcement learning-based system to dynamically adapt the robot navigation system based on longtime changes made in the environment. The targeted service environments are hotels, hospitals, houses, museums, offices, etc. The proposed navigation framework is generic and caters to the needs of various service environments.

The use of LfD techniques to teach robots to perform tasks has gained more traction in recent years. In LfD, a human operator teaches a robot to do a particular task, and the robot learns and repeats the same. It has numerous advantages when compared to the traditional hard coding technique. It is user-friendly, and even a novice user can teach a robot to do a task without needing to be a programmer to do it. If LfD is used, then the robot can be ported to any environment as the only prerequisite is to teach the robot, whereas, in traditional methods, the robot has to be reprogrammed. Most of the work done in this domain is on manipulators, and minimal work is focused on mobile robots (Ravichandar et al., 2020). The LfD techniques are implemented and tested in the 2D robotic simulator Player/Stage and an indigenously built mobile robot platform AAMoR. The system accuracy was tested in the AAMoR platform. A human operator has taught the robot a particular trajectory. The robot was able to replicate the trajectory autonomously and accurately.

Path planning is an active research area with much scope for advancement. Path planning plays a vital role in a mobile robot navigation system. It generates the shortest traversable path between two given points. Two major classes of path-planning algorithms are based on environmental awareness, fully known and unknown (Nakhaeinia et al., 2011). The former class path planner uses information from maps or environment sensors (e.g., beacons, cameras) to plan the path, which will restrict further environmental changes. In contrast, the latter class path planner is based on exploration without prior knowledge of the environment, which would lead to an inefficient path and take more time to explore the environment. The LfD system and TBP path planner are implemented and tested using the Player/Stage simulator to train the partial number of paths in three environments: house layout, hospital layout, and a benchmark maze layout. The test results prove the proposed path planner has scalability, trainability, accuracy, and repeatability. The path planner is compared with various classical path planning algorithms, and the results show that the proposed path planning algorithm is on par with the other algorithms in terms of accuracy and efficient path generation. The efficiency of the state-based obstacle avoidance system is tested successfully using different form factors of obstacles using the Player/Stage simulation environment and AAMoR mobile robot platform.

A mobile service robot operates in a dynamic environment with other robots and human beings. The service environment is often considered large and unknown, and the robot is expected to operate lifelong. Some of the features of the service environment are unique, such as vast square footage, dynamic in terms of moving obstacles (robots and humans), prone to environmental changes, such as generating new routes, blocking old routes, and inability to augment (such as adding extra sensors) the environment. One of the path planner's main concerns is environmental changes, which will hinder its performance. The static path planning system that directly uses maps or explores an unknown environment will work fine if the environment does not change over time. It uses the initial environment configuration to identify efficient paths, and this configuration will not be valid if the environment changes. Hence, a path planner that dynamically plans the path based on the current configuration of the environment is necessary to tackle this problem. This type of path planner is classified as a dynamic path planner, which considers environmental changes throughout its lifetime. A navigation system with a dynamic path planner continuously monitors environmental changes and updates itself to provide efficient paths.

Considering these facts, it is evident that the path planning system for a service robot should be an artificial intelligence-based system that will be able to plan the path given less or no information about the environment and cope with the environment's changes over time. Lifelong learning is a factor that is considered to keep any artificial intelligent agent learning and evolving continuously based on the various changes in the environment over a long period (Thrun, 1995). There have been tremendous advancements in machine learning and artificial intelligence techniques, which can be leveraged to incorporate into a path planner and make it dynamic. One such type of machine learning algorithm is Reinforcement Learning (RL), which uses the Markov Decision Process (MDP) to learn to operate in an environment (Sutton and Barto, 1998). The critical difference between reinforcement learning and other machine learning algorithms is that reinforcement learning does not need a vast labeled data set to train the system. A path planning problem lacks a huge dataset and can be defined as an MDP-based problem. Hence, an RL based path planning system will make the overall navigation system dynamic, which is required for a mobile service robot. The overall navigation framework with the DQL agent is implemented and tested using the ROS framework with Turtlebot3 mobile robot in the Gazebo simulator. The experiment results show that the proposed RL system learns all the routes based on the initial topological map of three different service environments with an accuracy of over 98%. The efficiency of the proposed TL algorithm is evaluated in a modified environment. A comparative analysis of the proposed TL and non-TL agents is performed based on various evaluation metrics. The convergence time required for the TL agent is half that of the non-TL agent.

1.1 Problem Statement

This thesis aims to develop a generic navigation framework for indoor mobile service robots using learning from demonstration and reinforcement learning techniques. The system is dynamic, user-friendly, environment-independent, and scalable.

Key Assumptions

- The service environment is indoor and slippage-free. As most of the service environments are indoors, the same is considered. Wheel slippage is not considered as it is not the focus of the work.

- The user-demonstrated path is the most efficient. A human demonstration will always be to the shortest path to a destination because demonstrating the longest path would be time-consuming and inefficient.

- At least one path to all the possible destinations is demonstrated. It would be time-consuming for the user to demonstrate multiple paths to a destination.

1.1.1 Objectives

The objective of the research work is to design and develop a navigation framework that will be able to

- Achieve environmental awareness based on user training to traverse one path to each destination.
- Map, all the possible unknown logical paths to all the destinations, using learned paths.
- Perform path planning and navigation for the long-term and to cope with environmental changes over time.

1.2 Methodology

The LfD system and obstacle avoidance system are implemented and tested using Player/Stage 2D robotic simulator (Brian P. Gerkey and Howard, 2003) and real-time mobile robot platform Amrita Autonomous Mobile Robot (AAMoR). Player/Stage is an open-source 2D robotic simulator popular among the robotics research community. AAMoR is an indigenously built four-wheeled mobile robot with a differential drive mechanism powered by ARM7-LPC2148. It has three LV-MaxSonar - EZ1 ultrasonic sensors on the robot's front, left, and back sides. The path planning, lifelong learning, and overall navigation systems are implemented and tested in Player/Stage and Robotic Operating System (ROS) framework with Turtlebot3 mobile robot platform in the Gazebo simulator (Macenski et al., 2022; Koenig and Howard, 2004).

1.3 Major Research Contributions

This thesis aims to develop a navigation system considering the requirements and limitations of a service application. The major research contributions are listed below.

Navigation Framework for Mobile Service Robot

A learning-based navigation framework is proposed and implemented for a mobile service robot. The framework is designed and developed to cater to the unique needs of a service environment. The navigation system learns the environment based on user demonstrations. The system can dynamically adapt to changes in the environmental configuration. Three major subsystems of the navigation framework are Learning from Demonstration (LfD), Path Mapping and Planning, and Lifelong Learning.

Learning from Demonstration

For effective navigation, information about the environment is necessary. Obtaining environmental information is not possible every time. Instead, obtaining information related to the path of each possible destination is easy to obtain, even if the target environment is large and dynamic. This work contributes a user-friendly LfD technique that aims to learn to traverse from minimal user training using only the robot's onboard sensor data. Three LfD approaches are proposed. Each approach is incrementally derived based on the limitations of the previous approach.

Path Mapping and Planning

This major contribution is to automatically map all the learned logical paths and generate a path tree. The path mapping system identifies all the landmarks in the environment to map the logical paths. These landmarks and all the destinations are connected to form a path tree. The system also identifies reference points in the environment that can be used to auto-correct the position and orientation errors of the robot. The proposed path planner also avoids any dynamic or static obstacle detected while traversing a path autonomously.

Lifelong Learning

This contribution is inspired by the fact that most work in path planning considers the environment static. At the same time, this assumption can be valid in some cases but not in a service environment. This work proposes a RL based system to enable the path planner to adapt to environmental changes. After a successful validity analysis of the RL system using a Q-Learning-based agent, the proposed Deep Q-Learning (DQL) based system with a novel Transfer Learning (TL) algorithm is implemented and tested.

1.4 Publications

Journal Papers

- A. A. N. Kumaar and S. Kochuvila, "Mobile Service Robot Path Planning Using Deep Reinforcement Learning," in IEEE Access, vol. 11, pp. 100083-100096, 2023, doi: 10.1109/ACCESS.2023.3311519.
- Kumaar. A. A. Nippun, Kochuvila. Sreeja and Nagaraja. S. R.. "A Scalable Tree Based Path Planning for a Service Robot" Journal of Automation, Mobile Robotics and Intelligent Systems, vol.16, no.1, 2023, pp.31-45. https://doi.org/10.14313/JAMRIS/1-2022/4

- Nippun Kumaar, A. A., and T. S. B. Sudarshan. "Learning from Demonstration with State Based Obstacle Avoidance for Mobile Service Robots." Applied Mechanics and Materials, vol. 394, Trans Tech Publications, Ltd., Sept. 2013, pp. 448-455. Crossref, doi:10.4028/www.scientific.net/amm.394.448.

Conference Papers

- A. A. Nippun Kumaar and S. Kochuvila, "Reinforcement learning based path planning using a topological map for mobile service robot," 2023 IEEE International Conference on Electronics, Computing and Communication Technologies (CONECCT), Bangalore, India, 2023, pp. 1-6, doi: 10.1109/CONECCT57959.2023.10234766.
- Kumaar, A.A.N., Sudarshan, T.S.B. (2015). Path Mapping and Planning with Partially Known Paths Using Hierarchical State Machine for Service Robot. In: Kim, JH., Yang, W., Jo, J., Sincak, P., Myung, H. (eds) Robot Intelligence Technology and Applications 3. Advances in Intelligent Systems and Computing, vol 345. Springer, Cham. https://doi.org/10.1007/978-3-319-16841-8_5
- A.A. Nippun Kumaar, Sudarshan TSB, "Sensor Counter Approach for a Mobile Robot to Navigate a Path Using Programming by Demonstration," Procedia Engineering, Volume 30, 2012, Pages 554-561, ISSN 1877-7058, https://doi.org/10.1016/j.proeng.2012.01.898. (https://www.sciencedirect.com/science/article/pii/S1877705812009083)
- A. A. N. Kumaar and S. TSB, "Mobile Robot Programming by Demonstration," 2011 Fourth International Conference on Emerging Trends in Engineering & Technology, Port Louis, Mauritius, 2011, pp. 206-209, doi: 10.1109/ICETET.2011.30.

1.5 Organization of the Thesis

The remainder of the thesis is organized into four chapters as follows.

In Chapter 2, a critical review of the literature on all the building blocks of navigation is presented. The review was focused on studying the state-of-art in three areas: Learning from Demonstration, path planning, and Reinforcement Learning. The chapter concluded by listing the major research gaps identified, which were used to formulate the problem statement and objectives.

Chapter 3 outlines the overall proposed navigation framework. The navigation framework, with its various systems and interconnections, is described in this chapter. All the major systems are described as sections of the chapter. The proposed LfD techniques are presented in Section 3.2. This section is divided into three sub-sections describing the proposed three different LfD techniques. Each sub-section introduces the proposed technique with a detailed description using appropriate tools. The LfD section identifies the best LfD techniques to use in the navigation framework. The proposed path planner is presented in Section 3.3. The need and motivation of the path planner and the usage of the LfD system data as a seed date for the path planner are described in the preamble of the section. Overall, the section is divided into two sub-sections. Section 3.3.1 describes the proposed path mapping and planning system with an appropriate example case. Section 3.3.2 describes the proposed obstacle avoidance algorithm. Section 3.4 describes the proposed lifelong learning system using RL. The overall section is divided into three sub-sections. Section 3.4.1 starts by describing a simple Q-Learning-based framework to implement a path planner. A DQL framework is reported in Section 3.4.2. The importance of a dynamic path planner is outlined, and a detailed description of the proposed TL algorithm is provided.

The results and discussion of all the systems in the proposed navigation framework are described in Chapter 4. The results of the three proposed LfD techniques are presented in Section 4.1. In section 4.2, the results of path mapping and planning are discussed. Finally, in Section 4.3, the results and discussion of the lifelong learning system are discussed.

Chapter 5 summarizes the findings and concludes the thesis work. The areas left unexplored and new areas that have opened up as a consequence of this research work have been identified and discussed as future directions.

Chapter 2

State of the Art

The overall thesis work covers three main areas of focus. A critical literature review in the areas, Learning from Demonstration (LfD), path planning, and Reinforcement learning in path planning are carried out to study state of the art and identify the potential research gaps.

2.1 Learning from Demonstration

Learning from Demonstration (LfD) has become a central topic of robotics that spans general research areas such as human-robot interaction, machine learning, machine vision, and motor control (Halbert, 1984). Robot LfD started about 30 years ago and has grown significantly during the past decade. LfD takes inspiration from the learning abilities of humans through imitation, thereby developing methods by which new skills can be transmitted to a robot. LfD covers a broad range of applications. In industrial settings, the goal is to reduce the time and costs required to program the robot, where the rationale is to create, modify, or assemble a product rapidly without requiring help from an expert in robotics. As a first approach to LfD, symbolic reasoning was commonly adopted in robotics, with processes referred to as teach-in, guiding, or play-back methods (Dufay and Latombe, 1984). In these works, LfD is performed through a manual process. In LfD, a human operator teaches a robot to do a particular task, and the robot learns and repeats the same. LfD technique is user-friendly, and even a novice user can train the robot to perform a task rather than coding it.

Three main factors lead to using the LfD technique for robots (Argall et al., 2009) (Siciliano, 2008). The first factor is a technique to teach a robot; most researchers have been using imitation learning in which a human operator demonstrates the task the robot will imitate (Argall et al., 2009). The other method is a Playback technique, where the robot learns by storing some action variables during the learning phase. The robot uses these variables later to repeat the same task automatically. The second factor is the interface used for teaching, such as vision, data gloves, laser range finder, and kinesthetic teaching. Ohya (2002) states that to follow a human, a mobile robot needs to know the position of the person and must be able to determine its path to follow its target. In this study, a person is equipped with a light-emitting device, and the robot detects this device using a camera. The third factor is to generalize the task learned. Initially, generality was achieved by teaching the robot many times with various possible scenarios to make it dynamic and flexible. Recently, soft computing techniques like Artificial Neural Networks (ANNs), Radial-Basis Function Networks (RBFs), Fuzzy Logic, and Hidden Markov Models (HMMs) have been widely used (Liang et al., 2023; Li et al., 2023; Gamal et al., 2020; Xi et al., 2019; Duque et al., 2019).

One growth area of mobile robots is service robotics, and its potential application areas are domestic, office, hospitality, etc. These applications require high-level, complex programming design incorporating dynamic adaptability to the user environment. The goal of LfD is to provide an easy human-robot interactive platform. It is possible for trajectory learning to obtain robot control code automatically through Nonlinear Auto-Regressive Moving Average with eXogenous inputs (NARMAX) system identification method (Iglesias et al., 2007). A nonlinear autoregressive model models the relationship between the robot's sensor data and the actuator. Kerr et al. (2010) obtained the control code by observing a human demonstrator following the desired route and translating his behavior directly into code without programming. NARMAX modeling is used to train a mobile robot to navigate through a doorway by Iglesias et al. (2005), a U-shaped corridor by Nehmzow et al. (2007), and some complex trajectories by Iglesias et al. (2008). Ehrenmann et al. (2002) discusses a technique to train a mobile service robot with a manipulator. The training is carried out by observation, based on human demonstration via camera. Later, the recorded data are segmented, interpreted, and mapped to the robot. Tasks like holding cups and plates were demonstrated.` Alvarez-Santos et al. (2015) proposes a tour guide robot that learns a route by following the teacher and storing the route in terms of its pose. The authors proposed an onboard sensor-based localization technique, which helps the robot identify different locations and poses.

Algorithms for robot Learning from Demonstration seek to enable human users to expand the capabilities of robots through interactive teaching instead of explicit programming. One main focus was developing techniques that can efficiently learn from non-expert users. Chernova and Thomaz (2012) highlights the importance of designing algorithms and interactions with non-expert end-users in the loop. In this work, non-experts used and evaluated three different LfD algorithms in the same domain. Their results point out the non-trivial challenges of getting learning input from a non-expert human teacher. Grollman and Billard (2012) addresses an important topic for social LfD, learning from failed demonstrations. When demonstrations come from non-expert humans, it can be expected that not every demonstration will be successful. Their experiments show that failure data do not have useful information for the learner. Similarly, Baris et al. (2012) assumes that the user teacher is naive and proposes a unified LfD framework that supports three types of motion demonstration inputs - trajectory, keyframe, and hybrid - allowing users maximum flexibility in teaching new skills. The techniques for converting demonstration inputs into a keyframe representation and the resulting Keyframe-based LfD system perform on par with state-of-the-art methods while significantly broadening the range of user interactions.

The demonstration need not be exhaustive as it might not always be feasible. Some works focussed on deriving actions based on a few demonstrations and generalizing the demonstration that can be used in a similar but not the same scenario. Dong and Williams (2012) trains the robot with a library of activities from user demonstrations and uses it to recognize an action performed by an operator in real-time. The authors contribute a novel probabilistic flow tube representation that can intuitively capture a wide range of motions, a method for identifying the relevant features of a motion, and ensuring that the learned representation preserves these features in new and unforeseen situations. Knox et al. (2012) seeks to understand people teaching naturally and design learning algorithms that match or even enhance natural human tendencies. Here, the authors describe two experiments that examine how different conditions affect a human teacher's feedback frequency and the computational agent's learned performance. The results indicate that the robot's action selection may not only be limited to exploration or policy exploitation but could also be designed to, indirectly and subconsciously, alter the behavior of the human teacher. Jäkel et al. (2012) tackles learning high-level tasks instead of low-level motor skills by learning task models for dexterous manipulation tasks from observing human demonstrations. The learned task models use a feature space of automatically generated contact constraints used by a planning algorithm when executing the learned task. Yasser and Toyoaki (2012) alternately assumes that the task expert is unaware of their teacher's role. In their approach, the robot has to take the initiative and decide which observed actions the teacher performs should be learned. The authors present a novel fluid imitation engine that enables the learner to automatically segment and learn observed behaviors without the help of the demonstrator.

Besides training the robot, giving expert feedback on the robot's performance can speed up the convergence time. Brenna et al. (2012) looks at allowing a human operator to give advice and feedback on a learned policy during its execution using advice operators. This allows the operator to administer corrections to state action pairs during execution. Here, the implementation and experiments in the mobile robot motion control domain demonstrate the utility of learning with directed feedback versus giving more demonstrations of the task. In a different kind of high-level feedback, Mericli et al. (2012) looks at the scenario of a human demonstrator providing feedback to a robot learning motion control tasks (e.g., walking). Their novel twist is in using the human to learn a policy of when to pay attention to what. Through interaction with the user, the agent learns the appropriate resolution of the state to use at various points in the task to represent the optimal action in that state.

Osentoski et al. (2012) envisions a future of remote experimentation with robots learning from any human on the web. Their article details the design and implementation of the PR2 Remote

Lab that was used in the AAAI 2011 Learning by Demonstration challenge. The experience of the teams that used the remote lab during this event points out the important challenges and problems for the future of robot learning in the cloud. Pignat and Calinon (2017) proposes a human dressing assistant. The system encodes sensory information and motor commands to learn a skill. The hidden semi-Markov model is used to reproduce the learned skills. (Ghalamzan and Ragaglia, 2018) proposes a different approach to training a robot in a static environment and execution in a dynamic environment. The algorithm was tested using a manipulator.

A detailed comparison based on learning technique, teaching interface, trainer, learning sensor, type of learning, robot type, and application is tabulated in Table 2.1. Three teaching interfaces are mostly used: kinesthetic, remote control, passive, and active observation. Kinesthetic teaching is used in manipulators by guiding the manipulator physically to perform the task. Remote control teaching, called teleoperation, controls the robot using a remote interface. The robot can learn from observing the teacher's demonstration either passively without performing the task or actively performing the task. In most cases, humans are the teachers, but researchers have also tried using a robot to train another robot. There is a wide range of on-board and off-board sensors used in the robot for learning. Off-board sensors like cameras are used to observe the trainer and learn from the actions performed by the trainer. But for practical limitations, it is optimal to use on-board sensors for learning. Majorly two types of learning methods are proposed in the literature they are, playback and generalization. In playback, the robot simply imitates the trainer's actions, and in generalization, the robot learns the actions in a general way so that it can be used to perform an unknown action as well. These techniques have been implemented in two main types of robots: manipulators and mobile robots. In mobile robots very minimal work has been carried out to learn complete paths in an environment rather only specific functionality of the navigation system is learning like, doorway crossing, obstacle avoidance, corridor wall following, and single trajectory following.

2.2 Path Planning

This has been an active research area for the past few decades. Many techniques have been developed for path planning and efficient navigation systems for mobile robots. These techniques can be broadly classified into three categories based on the robot's awareness of the environment (Nakhaeinia et al., 2011). They are path planning with known, unknown, and partially known environments.

Path planning with a known environment is also called deliberative planning. In this category of algorithms, the aim is to search for an optimal path and to generate a plan to reach the goal. It's a top-down approach. It requires the environment map for path planning in advance. Classical heuristic search algorithms like Dijkstra's and A* are the foremost techniques in this category

Table 2.1: Learning from Demonstration techniques comparative table

Paper	Learning Technique	Teaching interface	Trainer	Learning sensor	Type of learning	Robot type	Application
(Dufay and Latombe, 1984)	Inductive Learning	Kinesthetic	Human	Joint encoders	Playback	Manipulator	Assembly tasks
(Liang et al., 2023)	Neural network	Remote control	Human	Joint encoders	Playback	Manipulator	Pick and place
(Li et al., 2023)	Gaussian mixture regression and Neural network	Passive observation	Human	Kinect	Playback	Mobile robot	Navigation
(Gamal et al., 2020)	Fuzzy system and Multimodal deep Learning	Remote Control	Human	Camera and Proximity	Generalization	Mobile robot	Indoor navigation
(Xi et al., 2019)	Hidden Markov model	Passive observation	Human	Kinect	Generalization	Manipulator	Simple movements
(Duque et al., 2019)	Gaussian mixture model	Passive observation	Human	Kinect	Generalization	Manipulator	Assembly operations

(Iglesias et al., 2007)	Nonlinear Auto-Regressive Moving Average with eXogenous inputs (NARMAX)	Passive observation	Mobile Robot	Camera	Generalization	Mobile robot	Wall following
(Kerr et al., 2010)	NARMAX	Passive observation	Human	Camera	Playback	Mobile robot	S and U shaped trajectories
(Iglesias et al., 2005)	NARMAX	Passive observation	Human	Camera	Playback	Mobile robot	Doorway entry
(Nehmzow et al., 2007)	NARMAX	Passive observation	Human	Camera	Playback	Mobile robot	Corridor wall following
(Iglesias et al., 2008)	NARMAX	Passive observation	Mobile Robot	Camera	Playback	Mobile robot	8 shape trajectory
(Ehrenmann et al., 2002)	Neural network and Semantic interpretation	Passive observation	Human	Data glove, Magnetic tracker and Camera	Generalization	Manipulator	Pick and place

(Alvarez-Santos et al., 2015)	Augmented Monte Carlo Localization Algorithm (AMCL)	Active observation	Human	Camera, 2D laser scanners, WiFi and Magnetic compass	Playback	Mobile robot	Tour guide robot
(Grollman and Billard, 2012)	Learning from failure	Kinesthetic	Human	Joint encoders	Playback	Manipulator	Object manipulation
(Baris et al., 2012)	Keyframe based LfD	Kinesthetic	Human	Joint encoders	Playback	Humanoid Robot	Manipulation
(Dong and Williams, 2012)	Probabilistic flow tube	Remote Control	Human	Joint encoders	Generalization	Manipulator	Pick and place
(Yasser and Toyoaki, 2012)	Fluid imitation	Passive observation	Mobile robot	Demonstrator action, perception and object state ion environment	Playback	Mobile robot	Navigation

(Jäkel et al., 2012)	Planning model	Passive observation	Human	tactile glove, Motion tracker and Stereo camera	Playback	Manipulator	Spoon lifting, grasping and Bottle opening
(Brenna et al., 2012)	Focussed feedback for mobile robot policy	Active observation and Feedback	Human	Camera	Playback	Mobile robot	Navigation
(Mericli et al., 2012)	Multi-resolute corrective demonstration	Remote control and Feedback	Human	Joint encoders and camera	Playback	Humanoid Robot	Obstacle avoidance
(Pignat and Calinon, 2017)	Hidden semi-Markov model	Kinesthetic	Human	Joint encoders	Generalization	Manipulator	Dressing assistance
(Ghalamzan and Ragaglia, 2018)	Incremental LfD	Kinesthetic	Human	Joint encoders	Generalization	Manipulator	Pick and place

(Dijkstra, 1959; Hart et al., 1968; Ferguson et al., 2005). Both techniques search for an optimum path between the start and endpoints. The notable difference is that the A* algorithm directs the search towards more optimal states to reduce complexity. Anytime Repairing A* algorithms typically construct an initial, possibly highly suboptimal, solution very quickly, then improve the quality of this solution while time permits (Likhachev et al., 2003). Davoodi (2017) proposes Bi-objective path planning. The two objectives are finding the shortest path and achieving the maximum distance from obstacles using a deterministic algorithm. In deliberative planning, the environment map should be known as apriori. This approach cannot be used in dynamic unknown environments, which is essential for service robots.

Path planning in an unknown environment is known as reactive planning. This technique overcomes the drawback of deliberative planning in an unknown dynamic environment. The robot uses onboard sensors to explore the environment and decide upon the action to be taken. In this category of algorithms, the path is planned on the fly depending on the environment. Various approaches like the Motor Schema-based approach, Potential Field method, Improved potential field method, artificial potential field method, Vector Field histogram, and Modified flexible vector field method are proposed (Arkin, 1989; Huq et al., 2008; Hwang and Ahuja, 1992; Lee et al., 2012; Vadakkepat et al., 2001; Borenstein and Koren, 1991; Hong et al., 2007).

Another sub-class of reactive planning explores the environment and records the environmental information in some form. Later, path planning is carried out based on the recorded information. Rashid et al. (2013) proposes a path-planning technique using a binary visibility tree. The algorithm constructs a set of all possible paths and maps a binary visibility tree based on environmental obstacles. Coverage Based Path Planning method helps determine a path that passes over all points of an area or volume of interest while avoiding obstacles (Galceran and Carreras, 2013). Rapidly exploring Random Tree (RRT) generates paths with local constraints using probability (M. Lavalle, 1999). Another work uses probability for its navigation function and Gaussian probability distribution for the location of obstacles and robots in the environment (Hacohen et al., 2017). Tahir et al. (2018) discusses the problem of convergence time in a cluttered environment by bi-directional RRT* and intelligent bidirectional RRT*. The authors propose a potentially guided bidirectional tree, which improves the convergence rate. A common assumption in these techniques is that the environment is unknown, which is one of the basic requirements for a service robot. However, the path may not be as optimum as in the case of deliberative planning techniques. Reactive planning is best suited for service environments but with the cost of exploration time, which can be more for bigger environments.

Path planning with a partially known environment is a combination of both deliberative and reactive planning techniques. This is a hybrid technique. Here, path planning is carried out using deliberative planning, and then reactive planning is used if it encounters any obstacle in the previously planned path. Some heuristic approaches like D* and Field D* belong to this category (Stentz, 1994; Ferguson and Stentz, 2005). D* algorithm is derived from the A* algorithm, which can dynamically re-plan the path if necessary. In other studies, global path planning algorithms that carry out path planning using A* deliberative planning technique and then using a reactive planning technique like conjugate gradient descent, improved potential field, or D* Lite to improve the quality of the path locally have been reported (Dolgov et al., 2008; Zhenjun et al., 2007; Xu et al., 2007). Anytime Dynamic A* algorithm combines both possibilities of planning anytime and replanning in a dynamic environment (Likhachev et al., 2005). This hybrid technique works for dynamic environments such as service environments. However, a prerequisite for this algorithm is the initial static map of the loaded environment, and paths planned using deliberative planning.

The path planning problem can be categorized into two classes of problems, global and local path planning (Liu et al., 2023). Global path planning is planning a path from a source to a destination, given the environment map. Local planning is planning the path if an unforeseen scenario (obstacle) occurs while traversing the global path. Extensive research is carried out on applying the latest advancements in soft computing techniques to robotic path planning. Many recent techniques that use soft computing techniques for path planning have been proposed, such as Evolutionary algorithms, genetic algorithms, and Neural Networks (Yan et al., 2007; Lei et al., 2006; Zhang et al., 2008; Rahmaniar and Rakhmania, 2021; Nair and Supriya, 2020). Neuro-fuzzy hybrid inference system (Pandey et al., 2016). The hybrid algorithms combine two soft computing techniques and are better for the path planning systems (Mac et al., 2016). Hank and Haddad (2016) presents path planning in a partially known environment. An offline path planner will initially plan the paths with the given environment. During actual execution, an online path planner uses reactive path planning for unknown obstacles. A fuzzy inference system is used for trajectory tracking. Adaptive particle swarm optimization techniques are used for path planning (Dewang et al., 2018). Objective functions are based on the distance between the robot to the goal and the robot to the obstacle. Path planning using genetic algorithm and adaptive fuzzy-logic control is proposed by Bakdi et al. (2017). Algabri et al. (2015) propose fuzzy logic controllers to navigate an unknown environment. A cloud-based map storage approach is discussed by Limosani et al. (2016) in which the complete environment information is stored in the cloud. Method of Evolving Junctions is an optimization algorithm to find an optimal path in a 2D environment with dynamic obstacles (Li et al., 2017).

A predictive algorithm to avoid dynamic and static obstacles using the range sensors in the robot is discussed by Kamil et al. (2017). The predictive algorithms predict the velocity vectors of the dynamic obstacles and plan a collision-free path toward the goal. Choi et al. (2017) discusses a nonparametric motion controller using Gaussian process regression for trajectory prediction. Predicted trajectories and the robot's limited perception are used for robot motion control.Wang (2019) discusses a SLAM algorithm based on particle filter optimization. Image processing-based navigation techniques are also explored extensively. A dynamic landmark identification system using the visual servo method is proposed by Wang et al. (2004).

Due to the advent of the popularity of mobile robots and their application in social settings, research focusing on socially aware navigation has gained attention in the last decade. The key research problems addressed are human-aware navigation and human-robot interaction (Gómez et al., 2013). Human-aware navigation incorporates human motion trajectory into consideration for path planning. Human trajectory prediction using a social force model in an unknown environment is proposed by Kivrak et al. (2021). To track and recognize humans in the environment, a multi-sensor fusion technique with unscented Kalman filters to track the humans accurately is proposed by Bellotto and Hu (2010). Delivery robots are expected to navigate outdoors along with human pedestrians. Pedestrian prediction using social and psychological constraints is discussed by Bera et al. (2017). The concept of hierarchical path planning with local crowd prediction using the robot's onboard sensors and global path planning with the help of crowd information is proposed by Chen et al. (2023). An inverse reinforcement learning-based path planner is proposed to train the model based on humans in the environment by Kim and Pineau (2016). Another approach uses inverse reinforcement learning and rapidly exploring random trees to address the problem of the vast difference between learned and executed paths proposed by Ding et al. (2023). While path planning based on the shortest distance and human awareness is important, it is equally important to consider other social norms like human personal zones and restricted areas. A path planning model incorporating such constraints is proposed by Vega et al. (2020).

A detailed comparison based on planning technique, Planning type, environmental awareness, sensors used, and dynamic obstacle avoidance support is tabulated in Table 2.2. It is evident that deliberative planning is used for global planning and reactive planning is used for local planning. Few techniques use either of the planning methods, while other techniques use both of these planning methods for efficient path planning. Apart from environmental awareness, few works have focussed on obstacles and their configuration as well. This is essential for the autonomous nature of the mobile robot. Knowledge of the obstacle location and the ability to avoid dynamic moving obstacles is very crucial for an efficient mobile robot navigation system. Few systems are aware of the static obstacle location, while others use onboard sensors to identify the obstacles on the fly.

2.3 Lifelong learning

A path planned by a navigation system is as good as its understanding of the environment. Initially, the main focus of the path-planning research community was to achieve optimal paths given different levels of environmental awareness. These works are categorized based on the environment information available to the path planners (Nakhaeinia et al., 2011). With the developments in artificial intelligence and machine learning domains, using these techniques to solve the path-planning problem has gained more attention in recent years. This research direction will develop dynamic path planners best suited for service environments. A significant step in this direction is using

Table 2.2: Path planning techniques comparative table

Paper	**Planning Technique**	**Planning Type**	**Global/Local planning**	**Static Obstacle Location**	**Sensors used**	**Dynamic Obstacle avoidance?**
(Dijkstra, 1959)	Dijkstra	Deliberative	Global	Known	None	No
(Hart et al., 1968)	A*	Deliberative	Global	Known	None	No
(Arkin, 1989)	Motor schema	Deliberative	Global	Known	None	Yes
(Huq et al., 2008)	Motor schema and Fuzzy Logic	Deliberative	Global/Local	Unknown	Proximity	Yes
(Lee et al., 2012)	New Potential functions with Random Force algorithm	Deliberative	Global/Local	Known	No	Yes
(Vadakkepat et al., 2001)	Evolutionary Artificial Potential Field	Deliberative	Global/Local	Known	Overhead Camera	Yes
(Hong et al., 2007)	Modified Flexible Vector Field Approach	Reactive	Local	Unknown	Laser range finder and IR	Yes
(Stentz, 1994)	D*	Deliberative or Reactive or Hybrid	Global/Local	Unknown	Proximity	Yes

(Ferguson and Stentz, 2005)	Field D*	Deliberative or Reactive or Hybrid	Global/Local	Unknown	Proximity	Yes
(Dolgov et al., 2008)	Practical search techniques	Deliberative	Global/Local	Unknown	LIDAR and RADAR	Yes
(Zhenjun et al., 2007)	A* and Improved Potential Field method	Hybrid	Global/Local	Known	None	Yes
(Xu et al., 2007)	A* and Improved D* lite	Hybrid	Global/Local	Known	None	Yes
(Rashid et al., 2013)	Visibility binary tree	Deliberative or-Reactive	Global/Local	Unknown	Proximity	Yes
(M. Lavalle, 1999)	Rapidly-Exploring Random Tree (RRT)	Reactive	Local	Unknown	None	No
(Yan et al., 2007)	Fast evolutionary algorithm	Deliberative	Global	Known	None	No
(Lei et al., 2006)	Improved Genetic algorithm	Reactive	Local	Unknown	Proximity	Yes
(Zhang et al., 2008)	Hybrid Genetic algorithm	Reactive	Global/Local	Known(Static Obstacle)	Proximity	Yes
(Likhachev et al., 2005)	Anytime Dynamic A*	Hybrid	Global/Local	Known	Proximity	Yes

(Likhachev et al., 2003)	Anytime A* with provable bounds on sub-optimality	Hybrid	Global/Local	Known	Proximity	Yes
(Hwang and Ahuja, 1992)	Potential Field	Deliberative	Global	Known	None	No
(Borenstein and Koren, 1991)	Vector Field Histogram	Hybrid	Global/Local	Known(Static Obstacle)	Proximity	Yes
(Davoodi, 2017)	Bi-objective path planning	Deliberative	Global	Known	None	No
(Li et al., 2017)	Method of evolving junctions	Hybrid	Global/Local	Known (Static Obstacle)	Proximity	Yes
(Bakdi et al., 2017)	Genetic algorithm and Adaptive fuzzy logic control	Reactive	Local	Unknown	Kinect	Yes
(Choi et al., 2017)	Realtime nonparametric reactive navigation	Reactive	Local	Unknown	Proximity	Yes
(Dewang et al., 2018)	Adaptive Particle Swarm Optimization	Deliberative	Global	Known	None	No
(Hacohen et al., 2017)	Probability Navigation	Deliberative	Global	Known	None	No

(Hank and Haddad, 2016)	Hybrid approach with Fuzzy control	Hybrid	Global/Local	Known (Static Obstacle)	Proximity	Yes
(Tahir et al., 2018)	Potentially guided bidirectionalized RRT*	Reactive	Local	Unknown	None	No
(Kamil et al., 2017)	Predictive method	Reactive	Local	Unknown	Range finder	Yes
(Limosani et al., 2016)	Cloud Robotics	Deliberative	Global	Known	None	No
(Wang et al., 2004)	Vision based control	Reactive	Local	Unknown	Camera	Yes
(Nair and Supriya, 2020)	Recurrent Neural Network	Deliberative	Global	Known	None	No
(Pandey et al., 2016)	Adaptive Neuro-Fuzzy inference system	Reactive	Local	Unknown	Ultrasonic range finder and IR	No
(Rahmaniar and Rakhmania, 2021)	Genetic algorithm	Deliberative	Global	Known	None	No
(Kivrak et al., 2021)	Collision Prediction based Social Force model	Reactive	Local	Unknown	Proximity	Yes
(Bellotto and Hu, 2010)	A bank of Unscented Kalman Filters	Reactive	Local	Unknown	Multi-sensor	Yes
(Bera et al., 2017)	SocioSence	Reactive	Local	Unknown	Camera	Yes

(Brito et al., 2020)	Model predictive contouring control	Hybrid	Global/Local	Known	Camera	Yes
(Chen et al., 2023)	Hierarchical path planning	Hybrid	Global/Local	Known	Camera and LiDAR	Yes
(Kim and Pineau, 2016)	Inverse Reinforcement Learning	Hybrid	Global/Local	Known	RGB depth camera	Yes
(Ding et al., 2023)	RRT* and Inverse Reinforcement Learning	Hybrid	Global/Local	Unknown	Proximity	Yes
(Vega et al., 2020)	Socially accepted path planning	Deliberative	Global	Known	None	No

Reinforcement Learning (RL) based techniques (Sutton and Barto, 1998). Many RL algorithms are used in various domains like game automation, robotics, information theory, and operations research. A novel Twin Delayed Deep Deterministic policy gradients (TD3) framework for indoor mobile robot path planning is proposed by Gao et al. (2020). A 2D environment is used to train the RL agent, and then the agent is retrained in an actual 3D environment, reducing the training time. A multi-robot dynamic environment path planning is proposed in (Wang et al., 2020). Krishna Lakshmanan et al. (2020) proposes a novel deep black RL technique for a reconfigurable robot to perform tasks like floor cleaning, surface painting, tiling, inspection, and maintenance.

Q-Learning and Deep Q-Learning (DQL) are among the most used RL algorithms in various fields. A Q-learning-based path planning approach using only the sensory information of the robot is discussed by Zamstein et al. (2006). Romero-Martí et al. (2016) presents a path planning and navigation system for a service robot. This work successfully implements and tests a service robot used in a university for performing an autonomous task using Q-learning. A global path planner using a path graph as the input for the Q-learning-based RL agent to solve the path planning task is proposed in (Gao et al., 2019). A comparison study of Q-Learning and its modified State-Action-Reward-State-Action (SARSA) is carried out for global path planning in (Sichkar, 2019).

A Deep Reinforcement Learning (DRL) algorithm for solving grid path planning is studied in work (Panov et al., 2018). The authors conclude that a DQL path planner is robust in learning to reach the desired goal. Other works based on grid path planning are proposed (Hu et al., 2021; Pan et al., 2022). A DQL model used for path planning using image data is proposed and implemented by Xin et al. (2017). Another approach to explore in an unknown environment is to use sensor values to train the RL agent (Balachandran et al., 2022; Rajakani et al., 2022; Buitrago-Martínez et al., 2013). A warehouse mobile robot path planning using DQL is presented by Balachandran et al. (2022). DQL based algorithm for dynamic environment path planning is proposed by Pennock et al. (2018); Li et al. (2020). A dynamic environment is considered to have moving obstacles, and the obstacle location is unknown. Pennock et al. (2018) proposes a solution to replan a global path while facing an obstacle using DQL agent to achieve path planning in an unfamiliar dynamic environment. A multi-robot path planner using DQL is implemented by Yang et al. (2020). The paper proposed a methodology for fast convergence of the RL agent using prior knowledge of the environment. Another work on multi-robot path planning is presented by Bae et al. (2019), which uses the Convolutional Neural network-based DQL algorithm to train the agent using an environment map image. Zhu and Zhang (2021) presents a review article that comprehensively compares most mobile robot navigation systems using the DRL framework. One critical factor in any machine-learning algorithm is convergence time, i.e., the time the agent takes to learn the task. There have been considerable efforts to reduce the convergence time of DRL based path-planning algorithms. (Gao et al., 2020) reduces the convergence time by pre-training the RL agent using a 2D simulator environment and later using this experience to train in a real-time 3D environment. A Q-learning system with general information about the goal and current states is proposed by

Konar et al. (2013). This knowledge is used efficiently while training to reduce the training time significantly.

Researchers have made numerous attempts to solve lifelong learning in various domains. In (Widmer and Kubat, 1996), a notion similar to lifelong learning called concept drift is described, and a solution is proposed to use previous experience to learn new similar concepts. Correspondence learning is yet another attempt to use old experience in a new task efficiently. Ruman and Guy (2022) proposes a correspondence learning technique to use the previously learned skills to solve new tasks in an Atari ping-pong game environment. Transfer Learning (TL) is a technique that is applied to evolve machine learning models toward lifelong learning. A TL algorithm aims to transfer the knowledge of a task from a source model to a target model. Sutton et al. (2007) uses RL and TL tracking in stationary environments. Key points to be considered while applying transfer knowledge are the source and target tasks similarity, mapping between source states and target states, and the type of knowledge to be transferred (Taylor and Stone, 2009; Zhu et al., 2023). Applying TL between the source and target tasks with different states and actions, the challenge lies in mapping the source states to target states. Torrey et al. (2008); Taylor et al. (2008) proposes a simple approach to manually provide the mappings as subject experts. While this is a simple solution, there can be practical difficulties in obtaining human experts to map all the applications. Inter-task mapping can be considered required or automatically learned to address this shortcoming. If no explicit mapping is required, the agent attempts to learn the abstraction of the Markov Decision Process (MDP), which are invariant even though actions and state variables change (Banerjee and Stone, 2007; Konidaris and Barto, 2007). Another approach is to learn inter-task mapping automatically using novel mapping learning methods (Kuhlmann and Stone, 2007; Talvitie and Singh, 2007). Applying heuristics in transfer learning is proven to have reasonable acceleration in target task learning. Bianchi et al. (2008) obtains heuristics by exploring the environment. Obtaining the heuristics as different cases based on source task learning and using it in target learning is proposed by Bianchi et al. (2015); Jr. et al. (2010).

A detailed comparison based on RL planning technique, state space, action space, reward type, environment awareness, and Obstacle Avoidance type is tabulated in Table 2.3. Most of the state space consider in the literature are the robots position, target position and environment configuration in terms of obstacle locations and map of the environment itself. Actions are considered to be the basic mobile robot movements. Majority of the work consider the action and state space as discrete. The obstacle location and environment awareness are also an important factors considered available in few literature.

2.4 Limitations and Research Gaps

The following research gaps are identified from the literature survey.

Table 2.3: RL techniques comparative table

Paper	RL Technique	State Space	Action Space	Reward Type	Environment Awareness	Dynamic Obstacle Avoidance?
(Gao et al., 2020)	Twin Delayed Deep Deterministic Policy Gradient (TD3) and Probabilistic Road Map (PRM)	Lidar data, angle & distance between agent and target point, linear velocity and angular velocity at previous time step	Linear and Angular velocity	Continuous	Unknown	Yes
(Wang et al., 2020)	Globally Guided RL (G2RL)	Location(Free cells, Static obstacles, and dynamic obstacles)	{Up, Down, Left, Right, Idle}	Discrete	Known	Yes
(Krishna Lakshmanan et al., 2020)	Deep Black RL	Shape, Cleaning status, and Environment map	{Up, Down, Left, Right, clockwise rotation, anti-clockwise rotation}	Continuous	Known	No
(Zamstein et al., 2006)	Q-Learning	Proximity sensor values	{Forward, Backward, Turn left, Turn right, Rotate left, Rotate right}	Discrete	Unknown	Yes

(Panov et al., 2018)	Deep Q-Learning	Nearby grid occupancy status	{Left, Right, Up, Down}	Continuous	Unknown	Yes
(Hu et al., 2021)	Q-Learning	Grid locations	{Up, Down, Left, Right}	Discrete	Known	No
(Romero-Martí et al., 2016)	Q-Learning	Each Location in Map	Moving from one location to another	Discrete	Known(Topological Map)	No
(Gao et al., 2019)	Q-Learning	Agent's Position	{Up, Down, Left, Right}	Discrete	Known(Path Graph)	No
(Sichkar, 2019)	Q-Learning	Grid locations	{Up, Down, Left, Right}	Discrete	Known	No
(Xin et al., 2017)	Deep Q-Learning	Image of the environment in agent's POV	{Turn left, Turn right, Forward}	Discrete	Unknown	Yes
(Balachandran et al., 2022)	Deep Q-Learning	LIDAR sensor data	{Straight, Slight right, Right, Slight left, Left}	Discrete	Unknown	Yes
(Pennock et al., 2018)	Double Deep Q-Learning	LIDAR sensor and Local target position	{Forward, Backward, Left, Right, Left front, Left rear, Right front, Right rear}	Discrete	Unknown	Yes

(Bae et al., 2019)	Deep Q-Learning	Environment Image	{Up, Down, Left, Right}	Discrete	Known	No
(Pan et al., 2022)	Each round path planning based on Q-Learning algorithm	Grid locations	{Up, Down, Left, Right}	Discrete	Known	No
(Li et al., 2020)	Deep Q-Learning	Grid Locations	Grid Locations	Discrete	Unknown	Yes
(Rajakani et al., 2022)	Q-Learning	LIDAR sensor and target direction	Velocity	Continuous	Unknown	Yes
(Buitrago-Martínez et al., 2013)	Q-Learning	Proximity sensor and Target distance	Move to goal, avoid obstacle, follow left wall, and follow right wall	Continuous	Unknown	Yes
(Konar et al., 2013)	Deterministic Improved Q-Learning	Grid locations	{Up, Down, Left, Right}	Discrete	Known	No

- LfD field has developed rapidly since its inception, but no general techniques can be used in different robotics domains. Most approaches are tested using a single domain with one or a few tasks on a robotic platform. This is the primary reason for the lack of comparison between algorithms. Most of the work done in LfD is on industrial manipulators. Using the same technique in mobile robots is impossible, as learning parameters and teaching approaches differ in both platforms.
- Existing LfD systems heavily depend upon off-board sensors like cameras and wearable sensors for the trainer. This is inappropriate for a service environment as augmenting the environment with additional sensors might not be feasible and cost-effective.
- In deliberative planning, the robot is assumed to have prior knowledge about its deployed environment. In a highly dynamic service environment, the assumption is invalidated, and obtaining prior knowledge of the environment may not be possible.
- In reactive planning, the environment is unknown to the robot. This would lead to a less optimal path and more time for environmental exploration. An optimal path planner for a service environment would know a few optimal paths apriori, which is better than knowing the whole environment and building all possible paths based on it. The newly built paths will be optimal based on known paths. This way, the environment is partially known regarding paths to certain locations.
- In the hybrid technique, the robot has to be pre-loaded with the environment map for planning the initial global route and then allowed to adapt to the changes in the environment. As discussed, pre-loading the map will not be feasible in a service environment.
- In most of the work carried out in RL-based path planning, the agent's state space is considered as images or sensory information. While this is a reasonable approach to define the state space for an RL agent, the shortcoming is that the state space will be bigger. The bigger the state space, the more time it takes for the agent to converge to optimal learning.

2.5 Conclusion

A detailed summary of the literature survey is discussed in this chapter, with the limitations and research gaps listed. One key finding is that there is seldom work focussed on navigation frameworks specific to service robots and their unique features. This thesis aims to develop a generalized LfD technique to train a mobile service robot to navigate an indoor service environment. The major features of the robot are environment portability, user-friendliness, and mobile robot platform independence. Environment portability can be achieved by training only with the onboard sensors available on the robot and not augmenting the environment. This makes the system independent of the environment. A simple, intuitive user interface without any wearable device for training makes it user-friendly, and even a novice user can use it. The system is generic if it can be used in any mobile robot platform in any indoor service environment with minimal modifications. A path planner for a service robot should plan efficient paths that must be derived from the learned path as a seed. Any navigation system is incomplete without an obstacle avoidance algorithm. A robust dynamic obstacle avoidance algorithm is developed to avoid the obstacles in the route and return to its desired path.

To enable the static path planner to adapt to environmental changes dynamically, it is converted to RL based path planner. The RL agent will be trained initially using the topological map of the environment. A topological map will have all the key information about the relationship between all the necessary locations in an environment. The path tree, which the static path planner generates, can be used as a topological map. For lifelong learning, a novel transfer learning technique that considers only state space difference is proposed in this work.

Chapter 3

Proposed Navigation Framework

This chapter describes the proposed navigation framework for mobile service robots. A mobile robot deployed in a service environment should be user-friendly, environment-independent, and able to operate in large dynamic environments. The proposed navigation system is designed and developed to cater to these major requirements of a service mobile robot.

3.1 Learning-Based Navigation Framework

A generic navigation framework for mobile service robots is proposed in this work. Figure 3.1 depicts the overall navigation framework.

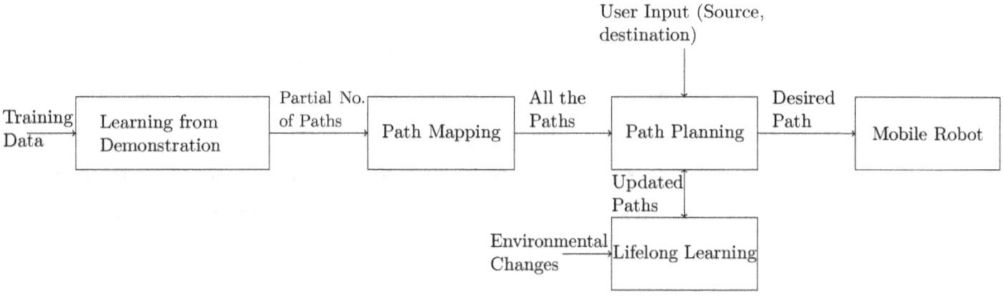

Figure 3.1: Proposed navigation framework

The initial focus of the navigation system is to obtain environmental information, which is achieved using the Learning from Demonstration (LfD) system. The robot traverses and collects training data based on the user's control. The robot is trained at least to traverse one path for all the possible destinations. This training information is processed and passed to the path mapping system that will map all the logical paths to all the destinations. The outcome in this stage is the path tree with all possible paths. The path planning module can use the path tree to plan the paths efficiently. For path planning, the robot's current location is considered the source location, and the user gives the destination location.

Since the service environment is prone to changes, the path tree has to be dynamically updated. The lifelong learning system uses a Reinforcement Learning (RL) agent to keep the path planner current with environmental changes. This will ensure that the path planner system always traverses the robot to a valid path to reach the desired destination. The chapter describes all the major systems of the navigation framework in detail.

3.2 Learning from Demonstration

The initial goal of a navigation system is to obtain environmental information for efficient path planning and maneuvering. An ideal scenario will be to obtain the full environment map, but in most applications, obtaining the map is difficult or impossible. Especially in a large dynamic service environment, it is not feasible to obtain the map. One of the major research goals is to design and develop a LfD technique to obtain environmental information based on human demonstrations. The benefit of using the LfD technique is that the robot learns the environmental information via user demonstrations rather than hardcoding it or obtaining it with extensive exploration. This chapter explains three proposed LfD techniques. These techniques differ in the learning process, and each technique is designed to overcome the drawbacks of its predecessor.

The main focus is to develop an efficient LfD technique that is user-friendly and environment-independent. The proposed LfD techniques consist of two phases: the teaching phase and the execution phase. In the teaching phase, the robot is trained to traverse all the possible destinations sequentially, starting from the home location with a minimal user interface. The home location can be considered as the charging point for the robot. Ensuring that the robot knows the environmental information in partial paths. In the execution phase, the robot will navigate autonomously with the taught information. A simple, user-friendly interface that a novice user can operate is used to control the robot while teaching. The robot's onboard sensor information is used to learn the path information. It does not depend on any environmental sensor, making the system environment independent and portable. As shown in Figure 3.2, a differential drive mobile robot is used for implementation and testing. The robot has three onboard sensors: wheel encoders, proximity, and pose sensors. Four proximity sensors are placed strategically, covering the robot fully.

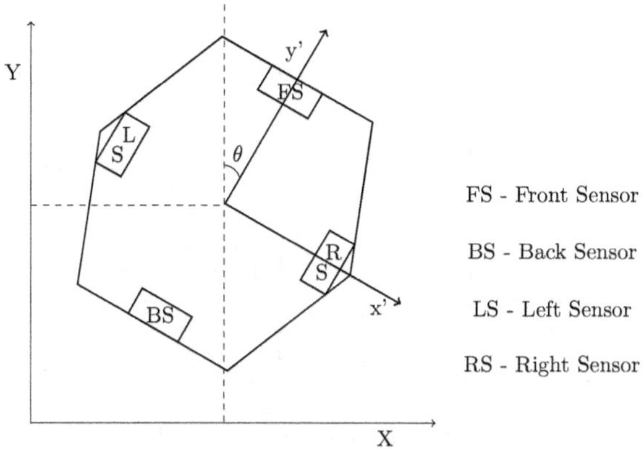

Figure 3.2: Robot's configuration and sensor location

3.2.1 Sensor Counter Based Learning Technique

This LfD technique uses a software counter and the robot's proximity sensor data to store the path information in a matrix data structure. The counter value is used to record the linear and angular distance of the robot. The counter is a time estimate of linear displacement in forward and backward directions and a function of angular displacement in turning left and right. The sensor value records the proximity information of the environment's static objects and configurations like furniture, hall, corridor, and doorway. In the teaching phase, both these matrices are updated periodically. In the execution phase, both counter and sensor values are used to navigate autonomously. The counter array for one path is given by Equation 3.1,

$$\text{CounterValue} = [D_1, C_1, D_2, C_2, D_3, C_3.......D_n, C_n] \tag{3.1}$$

Where D_i is the direction of movement encoded values of 1 to 4 for the front, back, left, and right, respectively; C_i is the software counter that counts the time of the robot's movement. While the robot starts to move, the software counter is started and runs until either the robot stops or the movement direction is changed. After the counter stops, the value is stored next to the direction index. This process repeats for every direction change. The array has been framed so that the even indices (0, 2, 4, 6, ...,n-1) are direction values, and all the odd indices (1, 3, 5, ...,n) are counter values of the immediate previous even value index. An example path and its data structure are depicted in Figure 3.3.

The distance information using proximity sensors is recorded with a constant sampling rate per path for only the forward movement. The sensor values are used to record low-level occupancy data while training. This data can be used to verify the accuracy of the autonomously executed

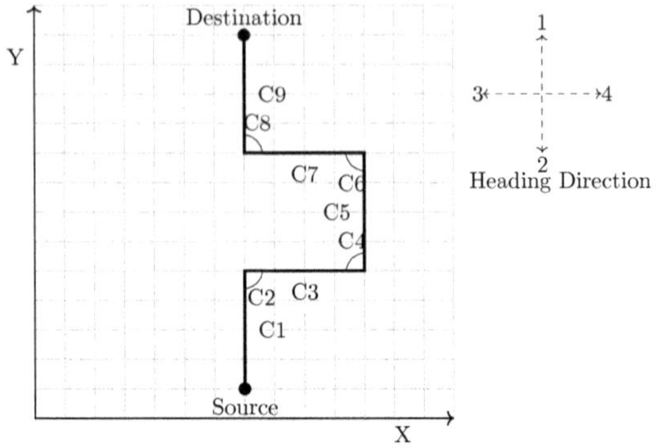

$$CounterValue = [1, C_1, 4, C_2, 1, C_3, 3, C_4, 1, C_5, 3, C_6, 1, C_7, 4, C_8, 1, C_9]$$

Figure 3.3: Example path with its counter values

path in execution mode. The sensor matrix with all the sensor values recorded for one forward segment in a path is given by,

$$\text{SensorValue} = \begin{bmatrix} V_{11} & V_{12} & V_{13} & V_{14} & \cdots & V_{1m} \\ V_{21} & V_{22} & V_{23} & V_{24} & \cdots & V_{2m} \\ V_{31} & V_{32} & V_{33} & V_{34} & \cdots & V_{3m} \\ \cdots & \cdots & \cdots & \cdots & \cdots & \cdots \\ \vdots & \vdots & \vdots & \vdots & \vdots & \vdots \\ V_{n1} & V_{n2} & V_{n3} & V_{n4} & \cdots & V_{nm} \end{bmatrix} \tag{3.2}$$

Each row in the matrix corresponds to data recorded at a time stamp, and its size n depends on the sampling rate. The first column contains the direction value; the remaining columns denote proximity sensor values. The matrix size is mxn, where m denotes the column size, which is determined by the total number of sensors and direction value ($m =$ Total No. of sensors+1) and n denotes the row size, which is determined based on the sampling rate at which the sensor data are recorded during training. For every path, all the forward segment sensor data are recorded in a separate sensor value matrix.

In the teaching phase, the robot learns the path by recording the direction, counter, and sensor values. These data are used to navigate the taught path autonomously. The recorded sensor value is used to verify if the robot is in the correct position along the path in the execution phase. Algorithm 3.1 and Algorithm 3.2 depict the counter-based technique algorithm for the teaching and execution phases, respectively.

Algorithm 3.1 Sensor Counter technique-Teaching phase

```
Initialize SoftwareCounter
Initialize CounterValue Array
Initialize SensorValue Matrix
Initialize I (User Input )
Initialize F (sensor value sampling frequency)
if I == 1 then
    Start SoftwareCounter
    Move forward
    if F then
        Update SensorValue
    end if
else if I == 2 then
    Start SoftwareCounter
    Move backward
    if F then
        Update SensorValue
    end if
else if I == 3 then
    Start SoftwareCounter
    Move left
    if F then
        Update SensorValue
    end if
else if I == 4 then
    Start SoftwareCounter
    Move Right
    if F then
        Update SensorValue
    end if
else if I == Stop then
    Stop moving
    Update CounterValue with I and SoftwareCounter
    Reset SoftwareCounter
    if F then
        Update SensorValue
    end if
end if
```

Algorithm 3.2 Sensor Counter technique-Execution phase

```
Initialize LocalCounter
CounterValue
L = Length(CounterValue)
for i=0;i<L;i=i+2 do
    if PathRow[i] == 1 then
        LocalCounter = CounterValue[i+1]
        while LocalConter != 0 do
            Move forward
            LocalCounter –
        end while
    else if PathRow[i] == 2 then
        LocalCounter = CounterValue[i+1]
        while LocalConter != 0 do
            LocalCounter –
            Move backward
        end while
    else if PathRow[i] == 3 then
        LocalCounter = CounterValue[i+1]
        while LocalConter != 0 do
            LocalCounter –
            Move left
        end while
    else if PathRow[i] == 4 then
        LocalCounter = CounterValue[i+1]
        while LocalConter != 0 do
            LocalCounter –
            Move right
        end while
    end if
end for
```

The software counter in this approach does not consider the robot's movement. To overcome this problem, Encoder Based Learning (EBL) technique is implemented;

3.2.2 Encoder Based Learning (EBL) Technique

This technique stores the wheel encoder pulse count instead of the counter value in the array. Sensor data, as in the sensor counter-based technique, is not used because, in a service environment, there is a high chance of symmetrical location. Two locations can be alike based on only the proximity sensor data. This technique records the encoder data of the learned path in a path matrix, as shown in Equation 3.3. Every row of the matrix is one path from the robot's home to a destination. The robot is trained to all the possible destinations from the home position. The home position is a point in the environment where the robot will be present if it is ideal. Most commonly, this is the charging point of the robot.

$$PathMatrix = \begin{bmatrix} D_{ij} & C_{ij} & D_{i(j+1)} & C_{i(j+1)} & \cdots & D_{i(j+n)} & C_{i(j+n)} \\ D_{(i+1)j} & C_{(i+1)j} & D_{(i+1)(j+1)} & C_{(i+1)(j+1)} & \cdots & D_{(i+1)(j+n)} & C_{(i+1)(j+n)} \\ \vdots & \vdots & \vdots & \vdots & \vdots & \vdots & \vdots \\ \cdots & \cdots & \cdots & \cdots & \cdots & \cdots & \cdots \\ \cdots & \cdots & \cdots & \cdots & \cdots & \cdots & \cdots \\ \vdots & \vdots & \vdots & \vdots & \vdots & \vdots & \vdots \\ D_{(i+m)j} & C_{(i+m)j} & D_{(i+m)(j+1)} & C_{(i+m)(j+1)} & \cdots & D_{(i+m)(j+n)} & C_{(i+m)(j+n)} \end{bmatrix} \tag{3.3}$$

In the path matrix, the number of rows in the matrix 'm' represents the total number of destinations, and the number of columns 'n' represents the various transitions the robot has to take before reaching the corresponding destination. The path variables are D and C, which are direction and distance/angle, respectively. The direction variable D is encoded as 1, 2, 3, and 4 for front, back, left, and right respectively. If the direction D is linear (front or back), then C represents distance L. If D is angular (left turn or right turn), then C represents the angle θ. The matrix is framed so that all the odd elements are direction values, and the immediate even elements are their corresponding distance/angle values. Every D and C is called a pose pair. This pair of data determines the pose in every state of the robot. Here, 'n' is the number of states the robot will transit through before reaching the corresponding destination. The 'n' is the maximum pose pairs a particular path consists of. If a path has fewer pose pairs than 'n', zeros are padded to the remaining pose pairs to keep the path matrix square. An example path and its path array are depicted in Figure 3.4 with nine pose pairs.

In the teaching phase, the robot records the pose pairs $(Direction, Displacement)$ for each path as row elements of the path matrix. The path matrix is used to navigate a path in the execution phase. This is depicted in Algorithm 3.3 and Algorithm 3.4.

The path matrix formed in the training phase is used to navigate autonomously. Unfortunately, these data are insufficient for a robot to reach a destination accurately and repeatedly. Mobile robots are prone to sensor errors, motor speed mismatch, and wheel misalignment, leading to lower accuracy and repeatability. Two major errors are identified: position error and orientation error. If position or orientation changes in the source or along the path, the destination point will vary

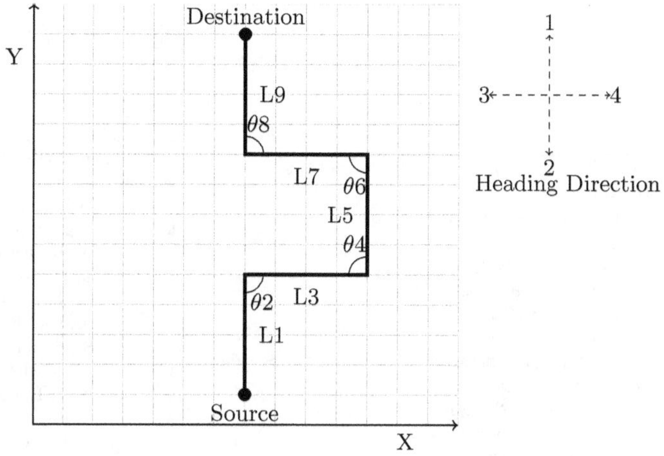

$$PathArray = [1, L_1, 4, \theta_2, 1, L_3, 3, \theta_4, 1, L_5, 3, \theta_6, 1, L_7, 4, \theta_8, 1, L_9]$$

Figure 3.4: Example path with its encoder values

relative to the error in position and orientation. Figure 3.5 and Figure 3.6 depict a scenario where the robot's starting position/orientation changes; the effect is the same if an error occurs along the path. As the path distance increases, the error also increases relatively. Training the robot to all the possible destinations from the home location is time-consuming for a larger environment with more destinations.

3.2.3 Enhanced Encoder Based Learning Technique

The Enhanced Encoder Based Learning (EEBL) technique is an upgraded version of the Encoder Based Learning (EBL) technique. Some key features of EEBL not in EBL are training to all the destinations sequentially, dynamic orientation, position error correction mechanism, and identification of reference points in the environment. Training is initiated from the home position to the first destination and continued sequentially till the path to the last destination is taught. This algorithm uses onboard sensors to learn paths taught by the trainer. The trainer uses a simple interface to move the robot front, left, right, and back. Wheel encoder data are stored in the path matrix to learn a path's linear or angular distance. Even though the training is carried out sequentially, the path matrix is always built based on the home position as a source. This is achieved by taking a partial path from the previous row in the matrix as per user input and appending the new path variables learned for a new destination. This is to maintain uniformity and easily process the path matrix for the next steps. When the robot navigates autonomously just with the distance information, it is prone to position and orientation errors.

Algorithm 3.3 Encoder Based Learning-Teaching phase

```
Initialize EncoderModule and EncoderValue
Initialize PathMatrix
Initialize Path Count M
Initialize I (User Input )
if I == 1 then
    Start EncoderModule
    Move forward
else if I == 2 then
    Start EncoderModule
    Move backward
else if I == 3 then
    Start EncoderModule
    Move left
else if I == 4 then
    Start EncoderModule
    Move Right
else if I == Stop then
    Stop moving
    Stop EncoderModule
    PathMatrix[M] ← (I,EncoderValue)
    Reset EncoderModule and EncoderValue
end if
```

Algorithm 3.4 Encoder Based Learning-Execution Phase

```
Initialize EncoderModule and EncoderValue
Initialize PathMatrix Row Index P
L = Length(PathMatrix[P])
for i=0;i<L;i=i+2 do
    if PathMatrix[P][i] == 1 then
        Start EncoderModule
        while EncoderValue ≤ PathMatrix[P][i+1] do
            Move forward
        end while
        Stop and Reset EncoderModule
    else if PathMatrix[P][i] == 2 then
        Start EncoderModule
        while EncoderValue ≤ PathMatrix[P][i+1] do
            Move backward
        end while
        Stop and Reset EncoderModule
    else if PathMatrix[P][i] == 3 then
        Start EncoderModule
        while EncoderValue ≤ PathMatrix[P][i+1] do
            Move left
        end while
        Stop and Reset EncoderModule
    else if PathMatrix[P][i] == 4 then
        Start EncoderModule
        while EncoderValue ≤ PathMatrix[P][i+1] do
            Move right
        end while
        Stop and Reset EncoderModule
    end if
end for
```

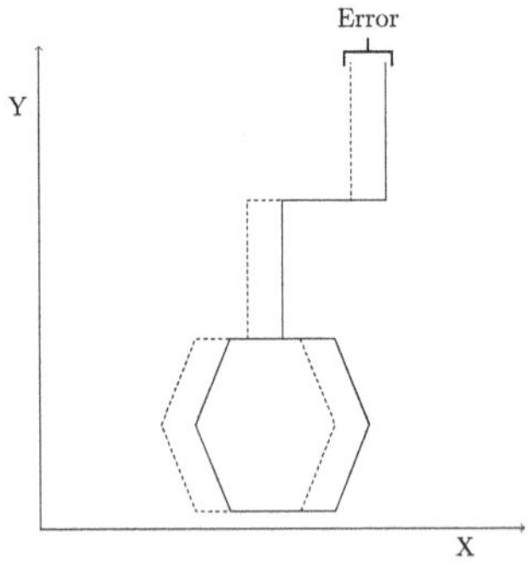

Figure 3.5: Displacement error (solid line without error and dotted line with error)

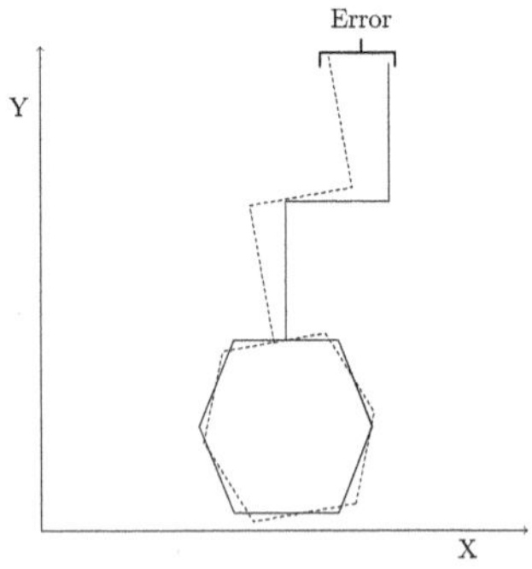

Figure 3.6: Orientation error (solid line without error and dotted line with error)

Position Error Position error occurs if there is a displacement in the robot's position. It is corrected by identifying the reference points in the environment. Reference points are fixed points in the environment, like walls and corridors. These points are identified using the robot's onboard proximity sensor. In the training phase, when the robot moves forward, it scans for reference points using its left and right proximity sensors. A location in an environment is selected as a reference point if the proximity value (either left or right sensor) is more than the given threshold value and is constant for a given sampling distance. When a reference point is identified, proximity values (left and right sensor) and its distance from the source location are stored in a reference matrix. Similar to the path matrix, each row of the reference matrix is for a path. There can be multiple reference points in a path. The reference matrix is updated in sync with the path matrix. Equation 3.4 is a reference array for one path with 'r' reference points. The array records the left sensor values (LSV) and right sensor values (RSV) for each reference point. To use the reference values in the execution phase, the path matrix is updated with the reference point location along a path segment. If the path segment consists of a reference point, then that path segment is represented as two pose pairs separated by a reference number. The reference number is used to access the LSV and RSV in the reference matrix.

$$ReferenceArray = [LSV_1, RSV_1, LSV_2, LSV_2, \ldots, LSV_r, RSV_r] \tag{3.4}$$

Orientation Error The orientation of a robot can change mainly due to hardware malfunctions like wheel misalignment and wheel encoder error. This causes overturning or under-turning of the robot, which will relatively change the destination pose. It can be corrected to an extent using the orientation information of the robot, which is obtained using a pose sensor. The pose sensor gives the robot's orientation angle θ about the Y-axis, as shown in Figure 3.2. In the training phase, orientation information is updated as pose angle θ in an orientation matrix at the beginning of each forward movement. This matrix is updated along with the path matrix and reference matrix. Equation 3.5 shows the orientation array for a single path.

$$Orientation\_array = [\theta_1, \theta_2, \theta_3, \ldots, \theta_p] \tag{3.5}$$

An example path segment with its path array, reference array, and orientation array is depicted in Figure 3.7. The reference and orientation matrices correct the position and orientation errors during autonomous navigation.

Error Correction Consider the example path shown in Figure 3.7. The grey blocks are walls, and the red dots along the walls are the points identified as reference points by the algorithm. Two factors must be fixed to find the reference point: proximity threshold and sampling distance. Let us consider the proximity threshold as 0.5 unit (the unit can be any standard unit of length)

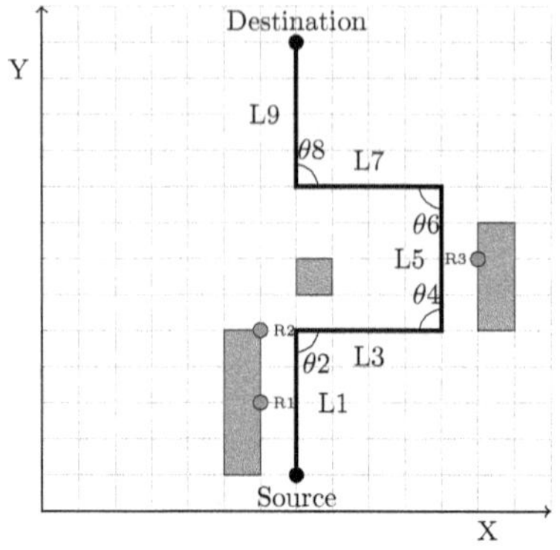

$$ReferenceArray = [LSV_1, Nil, LSV_2, Nil, Nil, RSV_3]$$

$$OrientaitonArray = [\alpha_1, \alpha_2, \alpha_3, \alpha_4, \alpha_5]$$

$$PathArray = [1, L_1/2, R1, 1, L_1, R2, 4, \theta_2, 1, L_3, 3, \theta_4, 1, L_5/2, R3, 1, L_5, 3, \theta_6, 1, L_7, 4, \theta_8, 1, L_9]$$

Figure 3.7: Example path with reference points (red circle) and its corresponding arrays (grey boxes are environment objects)

and the sampling distance as 1 unit. Assume the grids in the figure are 0.5x0.5 units. Based on these two parameters, the algorithm has identified three reference points (R1, R2, and R3) along the robot's forward movement. As soon as the reference point is identified, the path matrix is updated with the forward distance value until that point, and the corresponding sensor value is stored in the reference matrix. If only one sensor value satisfies the condition, the other values are stored as Nil. Orientation array is updated with pose angles θ_2, θ_4, θ_6 & θ_8 at the beginning of the forward movements L1, L3, L5, L7 & L9 respectively. In the autonomous phase, a PI controller auto-corrects the robot's position and orientation error. The identified reference point proximity values and pose angles are used as the reference values for position and orientation correction, respectively. The control equation is given by,

$$Y(t) = K_p e(t) + K_i \sum e(t) \qquad (3.6)$$

where,

e(t) = error (current_value - Reference_value)

Y(t) - controlled variable

Kp - proportional constant

Ki - integral constant

Algorithm 3.5 Enhanced Encoder Based Learning- Teaching Phase

```
Initialize EncoderModule and EncoderValue
Initialize PathMatrix, ReferenceMatrix, OrientationMatrix
Initialize Path Count M, Reference Index R
Initialize I (User Input )
if I == 1 then
    Start EncoderModule
    OrientationMatrix[M] ← (Current θ)
    Move forward
    if Reference Point then
        ReferenceMatrix[M, R] ← (Side Sensor Values)
        PathMatrix[P] ← (I, EncoderValue, R)
        Reset EncoderModule and EncoderValue
    end if
else if I == 2 then
    Start EncoderModule
    Move backward
else if I == 3 then
    Start EncoderModule
    Move left
else if I == 4 then
    Start EncoderModule
    Move Right
else if I == Stop then
    Stop moving
    Stop EncoderModule
    PathMatrix[M] ← (I,EncoderValue)
    Reset EncoderModule and EncoderValue
end if
```

Algorithm 3.6 Enhanced Encoder Based Learning- Execution Phase

```
Initialize EncoderModule and EncoderValue
Initialize PathMatrix Row Index P
L = Length(PathMatrix[P])
for i=0;i<L;i=i+1 do
    if PathMatrix[P][i] == 1 then
        Orientation Error Correction
        Start EncoderModule
        while EncoderValue ≤ PathMatrix[P][i+1] do
            Move forward
        end while
        Stop and Reset EncoderModule
        i++
    else if PathMatrix[P][i] == 2 then
        Start EncoderModule
        while EncoderValue ≤ PathMatrix[P][i+1] do
            Move backward
        end while
        Stop and Reset EncoderModule
        i++
    else if PathMatrix[P][i] == 3 then
        Start EncoderModule
        while EncoderValue ≤ PathMatrix[P][i+1] do
            Move left
        end while
        Stop and Reset EncoderModule
        i++
    else if PathMatrix[P][i] == 4 then
        Start EncoderModule
        while EncoderValue ≤ PathMatrix[P][i+1] do
            Move right
        end while
        Stop and Reset EncoderModule
        i++
    else if PathMatrix[P][i] == Reference Point then
        Position Error Correction
        Stop and Reset EncoderModule
    end if
end for
```

Here, the K_p value is assumed to be 1, and K_i is assumed to be 0.0001, obtained by trial and error. In the position correction system, the robot's position is corrected by controlling the sliding velocity using Y(t). In the orientation correction system, the robot's angular velocity is controlled using Y(t) to correct the robot's pose. Algorithm 3.5 shows the EEBL teaching phase with all possible robot movements and its matrices updating on different conditions for a path. Algorithm 3.6 shows the EEBL execution phase with the error correction mechanism in place.

Three proposed LfD techniques were described, and their results are discussed in Chapter 4. All three techniques used a simple user interface to train a robot's various paths in the teaching phase of the system. This makes the system user-friendly, where a novice user can train the robot. The main focus was learning the path's linear and angular displacements. In the sensor counter technique, the displacements were recorded using a software counter initiated for every linear or angular movement in the teaching phase. Even though the test results for a single path training were successful, there is a potential problem of false counting. This could happen if the robot is stuck for various reasons and does not move, but the software counter keeps incrementing. A simple solution for this problem is to use the absolute wheel rotation to calculate the linear and angular displacements. In the EBL technique, the software counter is replaced with a wheel encoder sensor to track the wheel movement accurately. Another major drawback is that the robot can be subject to position/orientation errors. Wrong starting position, sensor errors, motor speed mismatch, and wheel misalignment can cause this. The EEBL technique records data in path, reference, and orientation matrices to overcome this drawback. The reference matrix is updated based on the robot's side proximity sensor value at reference points like walls in an environment. These reference points are used to auto-adjust the robot's position in the execution phase to resolve position errors. The orientation matrix records the robot's orientation before every forward movement in the teaching phase. This matrix is used to overcome orientation errors. Hence, using EEBL technique, the mobile service robot is trained to navigate at least one path to all the possible destinations, which is the first objective of the work. These paths can be seed paths for the proposed navigation system's path mapping and planning sub-system.

3.3 Path Mapping and Planning

An essential input for any path planning system is its environmental information. The environmental information is obtained by simple user demonstration using the proposed LfD technique. At least one path to all the destinations in the environment is learned. There can be multiple ways to reach a point, but not all of them will be an efficient and shortest path. A path planner system is as good as its knowledge of its environment. The paths learned are partial. Not all possible paths are learned. This scenario is depicted in Figure 3.8. A path mapping system that converts the path matrix from the LfD system is converted into a path tree. The path planner module uses the path tree to plan the paths efficiently. While the robot is traversing a path autonomously, there can be obstacles. An obstacle avoidance algorithm is used to avoid the obstacle and return to the planned path. The proposed path mapping and planning system, along with an obstacle avoidance system, is described in this chapter.

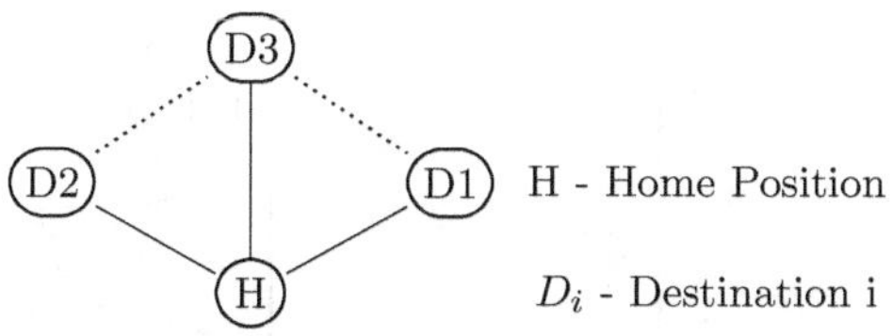

Figure 3.8: Illustration of path mapping (solid lines are learned paths, and dotted lines are unknown paths

3.3.1 Tree-Based Planner (TBP)

Tree-Based Planner (TBP) is a path mapping and planning algorithm which uses the learned path matrix as input. The path matrix formed by the EEBL algorithm has one possible route to all destinations from home, a partially known path. However, this information is not enough to plan efficient paths. Initially, using the TBP algorithm, all the other possible logical paths to all the destinations are mapped. As a result of path mapping, a path tree is formed with the home position as the root node and all the destinations as child nodes. Path mapping is carried out by identifying Virtual Land Mark (VLM)'s, which would act as a waypoint in the logical paths created and essentially map the shortest possible paths. VLM is identified based on the taught paths. An intuitive strategy is used to identify the VLM's. All the points where the robot takes a turn (left or right) are considered as VLM's. VLM's are the intermediate nodes in the tree. Later, given a source and destination, TBP uses the path tree to plan the efficient path to navigate autonomously.

Figure 3.9 depicts a scenario with a home position and three destinations, D1, D2, and D3. All the grey blocks are obstacles. The line between all the destinations is the path taught to navigate. The training sequence followed was Home $\rightarrow$ D4 , D4$\rightarrow$D1, D1$\rightarrow$D2, and D2$\rightarrow$D3, respectively. The trained path matrix is shown in Equation 3.7.

With this trained data, the Tree-Based Planner (TBP) algorithm generates VLM and forms the path tree. Two VLMs are identified (Black dot along the path in the figure), namely V1 and V2. V1 is chosen as the path has a left turn to reach D1. Likewise, V2 is chosen as there is a right turn to reach D2. The final path tree with VLM's and Destination is shown in Figure 3.10. Home position is the root node, all the VLMs are intermediate nodes, and all the destinations are child nodes. This is a ternary graph, where every node can have a maximum of three child nodes: the center, left, and right. These child nodes are symbolic representations of the direction of navigation from the parent to reach the child node. The robot must move from a parent node to reach a center child node. Similarly, left and right child nodes can be reached by taking appropriate turns. Every node contains information like forward distance, turning angle, reference point details, pose angle, and child node details.

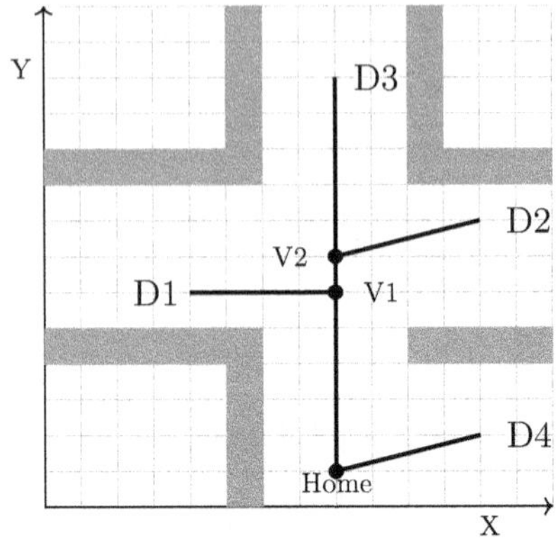

Figure 3.9: Sample multi-destination paths with identified VLM's (black circles)

$$SamplePathMatrix = \begin{bmatrix} 4 & \theta_3 & 1 & L_6 & 0 & 0 & 0 & 0 \\ 1 & L_1 & 3 & \theta_1 & 1 & L_2 & 0 & 0 \\ 1 & L_1 & 1 & L_3 & 4 & \theta_2 & 1 & L_4 \\ 1 & L_1 & 1 & L_3 & 1 & L_5 & 0 & 0 \end{bmatrix} \tag{3.7}$$

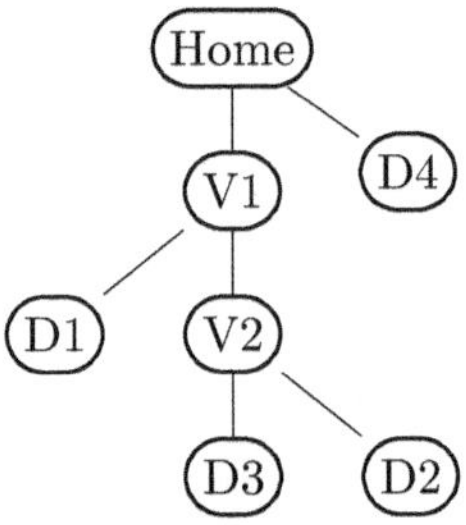

Figure 3.10: Tree generated by TBP algorithm with VLM's

TBP algorithm has two phases: path mapping and path planning phase. The path mapping phase has two stages: preprocessing and node formation. Algorithm 3.7 depicts the preprocessing stage. In this stage, the path matrix is split into three matrices based on the reaching destination: moving forward from home, moving left from home, and moving right from home. Since the root node of the tree is the home position. The splitting is based on the direction of the destinations from the home position. This would be the input for the node formation stage. The sample path matrix shown in Equation 3.7 is split into two submatrices, forward path matrix and right path matrix, as shown in Equation 3.8 and Equation 3.9.

$$SampleForwardPathMatrix = \begin{bmatrix} 1 & L_1 & 3 & \theta_1 & 1 & L_2 & 0 & 0 \\ 1 & L_1 & 1 & L_3 & 4 & \theta_2 & 1 & L_4 \\ 1 & L_1 & 1 & L_3 & 1 & L_5 & 0 & 0 \end{bmatrix} \tag{3.8}$$

$$SampleRightPathMatrix = \begin{bmatrix} 4 & \theta_3 & 1 & L_6 & 0 & 0 & 0 & 0 \end{bmatrix} \tag{3.9}$$

Algorithm 3.7 TBP algorithm- Path mapping- Preprocessing phase

for $ri = 0$; $i < max_r$; $ri++$ **do**
    for $rj = ri + 1$; $rj < max_r$; $rj++$ **do**
        if Path_Matrix[ri][0]>Path_Matrix[rj][0] **then**
            swap(ri, rj);
        else
            continue;
        end if
    end for
end for
int fcount, rcount, lcount;
for $r = 0$; $r < max_r$; $r++$ **do**
    if Path_Matrix[r][0] = = 1 **then**
        for $c = 0$; $c < mac_c$; $c++$ **do**
            Path_Matrixf[fcount][c]=
Path_Matrix[r][c];
        end for
    end if
    Similarly, for Left and Right directions, creating Path_Matrixl & Path_Matrixr matrix
end for

Algorithm 3.8 shows the node formation stage. In this stage, VLM's and their connectivity to home or any destination is identified. The outcome of this stage is a path tree with the home as a root node. .

Algorithm 3.9 shows the path planning phase. In the path planning phase, the formed path tree is used to navigate from a given source to a destination. In this phase, two paths are identified from home→source and home→destination. These two paths are merged on a common VLM node as a waypoint, and this would give the shortest route possible. If there is no common VLM, home is the common node that acts as a waypoint. This is the longest route possible.

Consider the tree in Figure 3.10. If a path must be planned between D3→D1, paths from home to both nodes must be identified. They are home→V1→D1 and home→V1→V2→D3. These two paths have V1 as a common node, merging the paths as V1 as a waypoint, we get

Algorithm 3.8 TBP algorithm- Path mapping- Node formation stage

```
fcount
if (fcount != 0) then
    for r = 0;r < max_r;r + + do
        for c = 0;c < max_c & Path_Matrixf[r][c]! = 0; do
            if Path_Matrixf[r][c] = = 1 then
                create-forward-VLM;
            end if
            Similarly for Left-VLM & Right-VLM
            c += 2;
        end for
        create-destination;
    end for
else
    Exit;
end if
```

Algorithm 3.9 TBP algorithm- Path planning

```
path1[m] = find_path(home to source)
path2[m] = find_path(home to destination)
for r = 0;r < max_r;r + + do
    for c = 0;c < max_c;c + + do
        if path1[r] == path2[c] then
            break;
        else
            continue;
        end if
    end for
end for
merge_routes(r,c)
```

D3→V2→V1→D1, this is the shortest logical path possible. The longest path without this technique would be D3→V2→V1→home→V1→D1 that goes through the home, which is unnecessary.

3.3.2 State Based Obstacle Avoidance Algorithm

The obstacle avoidance algorithm helps the robot overcome any obstacle. While the robot is traversing the path generated by TBP system, if any obstacle is detected, the robot has to overcome it and return to its original path. As the name suggests, this algorithm moves from one state to another based on the obstacle position and the robot's orientation.

As shown in Figure 3.11, there are five basic states: Diverge (D), Avoid (A), Converge (C), Diverge-Horizontal (DH), and Avoid-Horizontal (AH). Every state has three steps, each specifying the robot to move in a direction to avoid the obstacle. All turns taken in this algorithm are going to be 90 degrees. The states are:

Diverge (D): This state is executed if there is any obstacle in the robot path or the A state. In this state, the robot chooses the direction to diverge by the proximity values of the sensors on the side of the robot (left and right). Then, a turn is made in the chosen direction. It traces the side of the obstacle till the corner is reached. The distance covered while tracing is stored in terms of wheel encoder pulses in a variable d.

Avoid (A): This state is executed after D or AH states completion. Here, the robot is turned toward the direction opposite to the chosen direction of the last D state. For example, if the chosen direction is left in the last D state, the robot turns right in this state. Then, it avoids the gap between the robot and the corner of the obstacle. Then, it traces the side of the obstacle till its corner is reached.

Converge (C): This state helps the robot return to its original path. This path is executed after A state is completed. Initially, the robot turns toward the direction opposite to the chosen direction in the first D state. Then, it has to move forward to the distance d; finally, a turn to the same chosen direction as in the first D state has to be completed.

Diverge-Horizontal (DH): This state is activated if any obstacle is detected in the D or AH states. The robot chooses the direction to diverge by the proximity values of the sensors on the side of the robot (left and right). Then, a turn is made in the chosen direction. It traces the side of the obstacle till the corner is reached.

Avoide-Horizontal (AH): This state is activated after the DH state is completed. In this state, the robot is turned toward the direction opposite to the chosen direction of the recent D state. Then, it avoids the gap between the robot and the corner of the obstacle. This can be achieved by going forward till the corner is detected. Then, it traces the side of the obstacle till its corner is reached. The distance covered while gap avoidance and tracing are stored as wheel encoder pulses in the same variable d.

In Figure 3.12, the grey arrow shows the robot's path to be followed. An obstacle is found mid-way, then the state-based obstacle avoidance algorithm is triggered, and the basic three states, D, A, and C, help the robot avoid the obstacle and return to its original path.

A Tree-Based Planner (TBP) maps the partial paths learned using the EEBL system into a path tree. This path tree is used to plan an efficient path given the source and destination. On encountering an obstacle while traversing the path, the proposed state-based obstacle avoidance algorithm is used to avoid the obstacle efficiently and return to the path.

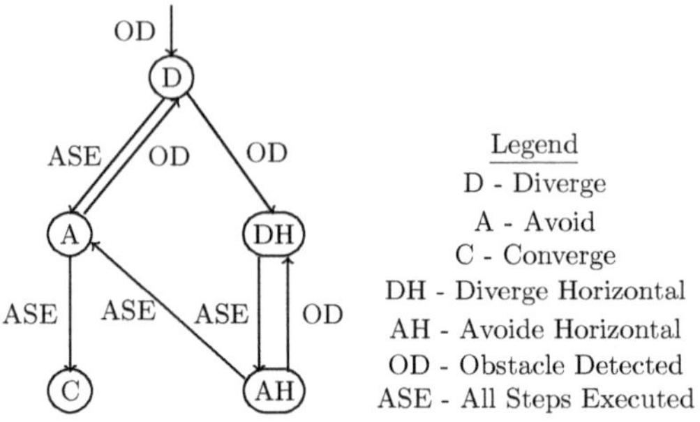

Figure 3.11: State diagram of state-based obstacle avoidance algorithm

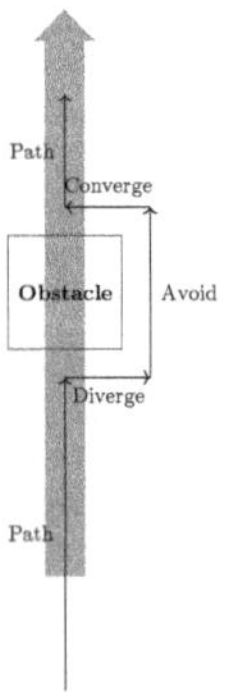

Figure 3.12: Simple obstacle avoidance scenario

3.4 Lifelong Learning

A service robot is expected to operate in an environment for a long period. The environmental configuration can change over time, and the path planner should be dynamic enough to adapt to the changes. An artificial intelligence-based dynamic path planner is proposed for lifelong learning using Reinforcement Learning (RL) and Transfer Learning (TL). The path tree from the TBP system is used to pre-train the RL agent. In the event of an environmental change, TL is used to learn the new changes and reuse the old training efficiently. The path planner based on RL and Deep Reinforcement Learning (DRL) with Transfer Learning (TL) is described in this section.

3.4.1 Q-Learning Based Path Planner

The proposed Reinforcement Learning (RL)-based path planning algorithm uses a Q-Learning algorithm to learn the paths using a topological map of the environment. The RL agent is trained initially using the topological map of the environment. A topological map contains all the key information about the relationship between all the necessary locations in an environment. The nodes in the map are the critical locations that are identified manually or using any other specialized techniques. The edges between the nodes represent the interconnection between the nodes. A service environment is typically considered dynamic and large in square footage. Generating a map of the environment is increasingly difficult, but a topological map with all the key locations for path planning can be generated. The path tree from the TBP system can be directly used as a topological map.

RL framework

In a general definition of a reinforcement learning system, an agent acts in an environment, and for every action, the agent gets rewarded positively or negatively. The agent aims to learn to act efficiently in the given environment for maximum positive cumulative reward. A Q-Learning-based RL algorithm to implement the path planner is proposed here. This is one of the most used RL algorithms to solve different problems. The parameters of the typical RL framework are State, Action, and Reward. States are considered all the finite configurations an RL agent can be at in an environment. Actions are the set of possible actions the agent takes to move from one state s to another state s'. Rewards are the feedback given to the agent for the last action taken. It can be a positive reward or a negative reward. This reinforces the agent to take actions that lead to positive rewards and avoid negative reward actions. The Q-learning agent trains to achieve the best Q values. Q-Values defined for states and actions $Q(s, a)$ estimate how good the action a is at the state s.

The Q-learning agent uses the Bellman equation to update the Q-values. It is shown in Equation

3.10, where α is the learning rate, a hyper-parameter to control the learning ability of the agent; $R(s,a)$ is the reward for the state action pair (s,a); $Q(s,a)$ is the current Q value acquired from Q-table; $maxQ'(s',a')$ is the maximum expected future reward based on the new state s' and all the possible actions; γ is the discount rate, a hyper-parameter controlling the weightage of the future reward.

$$NewQ(s,a) = Q(s,a) + \alpha[R(s,a) + \gamma maxQ'(s',a') - Q(s,a)] \tag{3.10}$$

States: States are represented as $s \in S$, where s is the robot's current state, a subset of all the possible states of an agent. In our work, S is the set of all the nodes in the topological map. At any given time, the agent, the robot, will be in one of the nodes, i.e., one of the locations in the environment.

Actions: Actions are represented as $a \in A$, where a is the action taken by the agent to reach the next state. As our agent is a mobile robot, A is the set of all the basic movements of the robot on a 2D plane, $A = \{left, right, forward, backward\}$.

Reward: Rewards are the state-action-based feedback given to the agent. The reward space $R(S,A)$ is the set of all the possible rewards based on the current state and action. Equation (3.11) shows the reward function of the proposed RL agent. While the value of the reward is constant, it can be changed as long as the difference between the different rewards is maintained. A positive reward motivates the agent to perform a similar action in the future, and a negative reward is to avoid the action in the future. The highest reward r_1 is to reach the goal state $g \in G$ where G is the set of all the goal states $G \subset S$. The highest negative reward r_4 is given for reaching an unknown state $U \not\subset S$. A reward r_3 is given to the agent to avoid looping. A reward r_2 is given as a step reward; this converges the agent to find the shortest path possible.

$$r(s,a) = \begin{cases} r_1, & s = g \\ r_2, & s = \{x \mid x \notin (S \cap G)\} \\ r_3, & \text{s is previously visited location} \\ r_4, & s = U \not\subset S \end{cases} \tag{3.11}$$

Learning process

The training is carried out as episodes. The agent takes multiple discrete steps in each episode. In each step, an action transits the agent's state. In this work, the states are considered as different physical locations in the environment, and actions are the basic movements of the robot. A reward is given for each such transition, and the agent updates the Q value based on the Bellman equation. An episode is terminated based on the current state s of the agent. Equation 3.12 depicts all the terminating conditions. The RL agent's goal is to get more rewards for each episode.

$$s = \begin{cases} \text{terminate\_state}, & s = g \\ \text{terminate\_state}, & s = U \not\subset S \\ \text{terminate\_state}, & \text{path length} == \text{total states} \\ \text{non terminal\_state}, & \text{Otherwise} \end{cases} \tag{3.12}$$

During training, the agent uses the next state table and Q-table to learn to reach the desired goal efficiently. The next state table is a look-up table the agent uses to transit between states given the current state action pair. The table consists of all the states as row elements and all the actions as column elements. The components of the table are the next state, given the state action pair. Q-table is used to update the Q-values for a particular state action pair. This table is used to evaluate the Bellman equation. Table 3.1 depicts the next state table and Table 3.2 depicts the Q-table for the sample path tree shown in Figure 3.10.

An important training parameter is ϵ, which decides the exploration and exploitation strategy that makes or breaks the agent training process. ϵ is the hyper-parameter controlling the agent to explore vs. exploit. During training, the agent chooses the action based on Q-values or a random action based on the ϵ value. The training algorithm is shown in Algorithm 3.10.

Table 3.1: Sample next state table

States	Forward	Left	Right	Backward
Home	V1	U	D4	U
V1	V2	D1	U	Home
V2	D3	U	D2	V1
D1	U	U	U	V1
D2	U	U	U	V2
D3	U	U	U	V2
D4	U	U	U	Home
U	U	U	U	U

Table 3.2: Sample initial Q-table

Home	V1	V2	D1	D2	D3	D4	U
V1	0	0	0	0	0	0	0
V2	0	0	0	0	0	0	0
D1	0	0	0	0	0	0	0
D2	0	0	0	0	0	0	0
D3	0	0	0	0	0	0	0
D4	0	0	0	0	0	0	0
U	0	0	0	0	0	0	0

Algorithm 3.10 Q-Learning training

Initialize Q_Table and Next_State_Table
Initialize Hyper-parameters - α, γ, ϵ, episode_count
for $episode <= episode\_count$ **do**
    $s = random(S)$
    while $s! = terminate\_state$ **do**
        if $random(0-1) < \epsilon$ **then**
            a $= a_{max}(Q\_value(s))$
        else
            $a = randint(1-4)$
        end if
        $s' = Next\_State\_Table(s, a)$
        calculate r(s, a)
        $New\_Q(s, a) = Q(s, a) + \alpha[r(s, a) + \gamma maxQ'(s', a') - Q(s, a)]$
        $Q\_Table(s, a) = New\_Q(s, a)$
    end while
end for

The proposed Q-Learning-based path planner was successfully modeled and tested for two environments. Based on the number of nodes and destination location, the hospital environment is about 3.5 times larger than the house environment. The results are discussed in Chapter 4, where the accuracy of both environments is above 95%

3.4.2 Deep Q-Learning (DQL) Based Path Planner

A typical service environment is vast, with many locations. The Q-Learning-based path planner uses a Q-table to find the best state-action pair, and the table size is proportionally increased based on the state space dimension. This makes the algorithm computationally expensive and requires more memory to store the Q-values. On the other hand, Deep Q-Learning (DQL) replaces the Q-table with a neural network capable of learning from a very large state and action space. The neural network estimates the Q-value given a state-action pair efficiently. Hence, a Deep Reinforcement Learning (DRL) based path planner will be more suitable for the service environment. The overall system consists of two modules: DRL module that uses DQL technique and the TL module. DRL module is used to train an RL agent with the initial environment configuration to plan the paths efficiently. TL module is used to re-train the RL agent if there is any configuration change in the

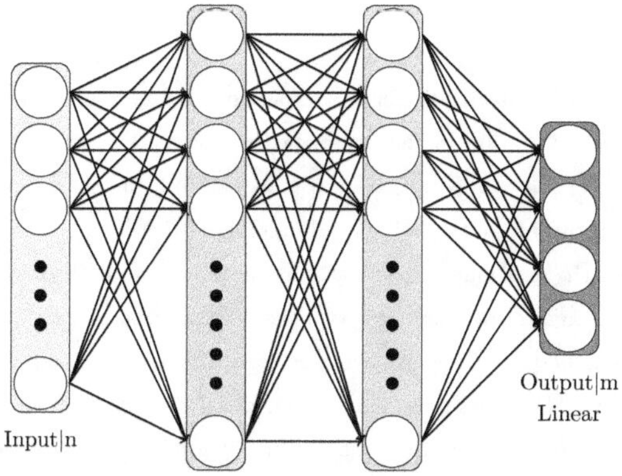

Figure 3.13: Proposed DQN architecture

environment. DQL uses a neural network to predict the best action given the current state. The parameters of the DQL framework are initialized based on the topological map used to train the RL agent. The states, actions, and rewards for the RL agent are the same as in Section 3.4.1.

Deep Q-Network Architecture

The Deep Q-Network (DQN) architecture consists of an input layer, two hidden layers, and an output layer with Adam optimizer, as shown in Figure 3.13. The neural network configurations depend on the states and actions of the RL agent, as shown in Equation 3.13 and Equation 3.14. The total number of input neurons n depends on the number of states S and the number of goals G, i.e., the input to the neural network depends on the current state s and the desired goal to reach g.The goal states $G \subset S$ denotes all the possible destination locations obtained from the topological map. Adding goal states as part of the input layer enables the agent to find the optimal path to all the possible destinations. The output neurons m depend on the number of actions A. This represents Q-values for all the actions based on the current state. The best action is based on the output layer's maximum Q-value.

$$n = |S| + |G| \tag{3.13}$$

$$m = |A| \tag{3.14}$$

Training

The neural network is trained using an experience replay buffer. Experience $e = (s, a, s', r)$ is stored after every iteration in the experience memory. A random experience batch is used for training. Batch size η is a hyper-parameter. Equation 3.15 shows the loss equation, which is the Bellman error derived from the Bellman equation for Q-Learning. $Q^*(s, a)$ is the predicted Q value. $maxQ'(s', a')$ is the maximum expected future reward based on the new state s' and all the possible actions. An RL agent at the early stages of the training process should explore more to understand the environment better. After exploring enough, the agent should rely on the trained network, i.e., exploitation. The hyper-parameter ϵ manages the exploration and exploitation factor. To set the ϵ value high to explore initially and reduce it gradually to exploit, the ϵ-decay algorithm is used. The decay function in Equation 3.16 is chosen to be an exponential function based on the ϵ, current training episode number, ϵ_min, ϵ_initial. The algorithm gradually reduces (decays) the ϵ value from ϵ_initial to ϵ_min. A training episode is terminated based on the current state s of the agent. Equation 3.12 in Section 3.4.1 depicts all the terminating conditions.

$$Loss = (Q^*(s, a) - (r + \gamma maxQ'(s', a')))^2 | e = (s, a, s', r) \tag{3.15}$$

$$\begin{aligned} \epsilon_{new} &= \epsilon_{old} \cdot e^{current\_episode} \\ where, e &= total\_episodes \sqrt{\frac{\epsilon\_min}{\epsilon\_initial}} \end{aligned} \tag{3.16}$$

β-decay Lifelong Learning Algorithm

The proposed β-decay algorithm is a transfer learning algorithm applied when a configuration change in the environment affects the trained DQL path planner. The non-TL approach would be to retrain the new RL agent from scratch. The configuration changes are less likely to be major and will be gradual. This will affect the number of nodes in the topological tree, i.e., there will be only a change in the state space of the RL agent. Considering this fact, the non-TL approach would be redundant and time-consuming as the old agent will have useful information that can be used to upgrade the new agent to incorporate the new environment configuration.TL approach is used to re-train the new agent using the knowledge of the old agent, hence reducing training time. The old agent path planner is considered the source task, and the new agent path planner is the target task.

The proposed β-decay TL algorithm transfers the knowledge of the source task to the target task based on the β factor. The target task agent learns from the source task knowledge or ϵ-decay learning process described in Section 3.4.2.

The higher the β value, the higher the possibility of transferring the source task knowledge. β is initialized based on Equation 3.17. The β is initialized based on the probability of choosing a random state from the target state space that is the same as the source state space. S_{target} is the state space of the target task and S_{source} is the state space of the source task. The decay factor is calculated as in Equation 3.16. Based on the ϵ and β, the proposed TL algorithm chooses between Experience vs. Exploration vs. Exploitation (EEE) as shown in Algorithm 3.11.

$$\beta = \frac{|S_{target} \cap S_{source}|}{|S_{target}|} \tag{3.17}$$

Algorithm 3.11 Experience vs. Exploration vs. Exploitation algorithm

Initialize Source and Target DQN network
Initialize parameters - ϵ, β, episode_count
for $episode <= episode\_count$ **do**
    $s = random(S)$
    while $s! = terminate\_state$ **do**
        if $random(0-1) < \beta$ **then**
            $Q\_value(s_{source}) = Source.Predict(s_{source})$
            $a = a_{max}(Q\_value(s_{source}))$
        else if $random(0-1) < \epsilon$ **then**
            $Q\_value(s_{target}) = Target.Predict(s_{target})$
            $a = a_{max}(Q\_value(s_{target}))$
        else
            $a = randint(1-4)$
        end if
    end while
end for

3.5 Conclusion

The overall navigation framework consists of three major systems in this chapter. First, the Learning from Demonstration (LfD) system uses the proposed Enhanced Encoder Based Learning (EEBL) technique to learn the partial number of paths from user demonstrations. This system makes the navigation framework user-friendly by keeping the training interface simple so a novice can train the robot and environment independent, as it uses only an onboard sensor to learn the paths. Second, the proposed Tree-Based Planner (TBP) maps all the possible logical paths based on EEBL system's output. Finally, the proposed Reinforcement Learning (RL) based path planner can dynamically adapt to the changes in the environmental configuration.

Chapter 4

Results and Discussion

All the major systems of the proposed navigation framework, Learning from Demonstration (LfD), path mapping & planning, and lifelong learning are implemented and tested. This chapter describes the detailed results and analysis of all the systems.

4.1 Learning from Demonstration

Three proposed LfD techniques are tested in a similar environment using the Player/Stage 2D robotic simulator. A wheeled mobile robot is simulated in a house layout environment as shown in Figure 4.1. The environment was chosen as a typical house layout with eight destinations (D1, D2, D3, D4, D5, D6, D7, and D8) and a home position. The home position can be considered the robot's charging point. The training starts from home position. When the robot is idle in the execution phase, it comes to the home position.

4.1.1 Sensor Counter Based Learning Technique

In Player/Stage, a path has been taught to reach the D3 of the house layout in the teaching phase, and the robot executes the same path automatically in the execution phase. As shown in Figure 4.2 in the execution phase, the robot follows exactly as learned in the teaching phase. Similarly, two more paths are trained to reach D4 and D5 from home. Figure 4.3 and Figure 4.4 depict the resulting trained and execution paths. The algorithm's performance is calculated by finding the absolute error difference between the robot's destination pose in the teaching and execution phases. The pose of any mobile robot constitutes position values concerning the X and Y axis and orientation concerning the Z axis. The pose is represented as [X, Y, Deg]. Table 4.7 tabulates all

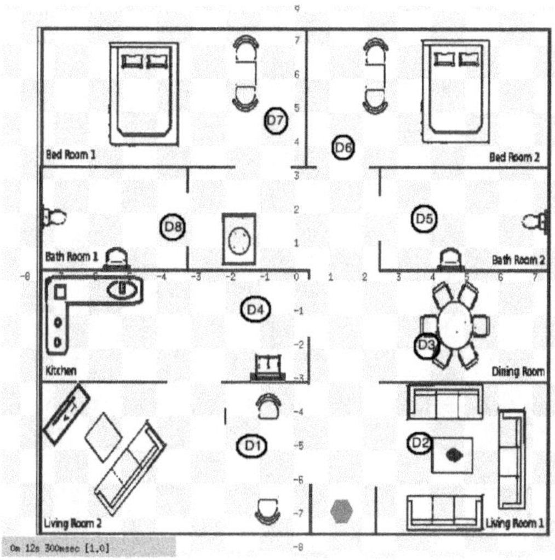

Figure 4.1: Player/Stage house layout environment with a robot in the home position (red hexagon-shaped object in the middle of the bottom side), marked destinations, and the world coordinate frame in the center of the environment

three paths of teaching and execution of final positions and their corresponding errors. This shows that the proposed technique efficiently executes the trained paths.

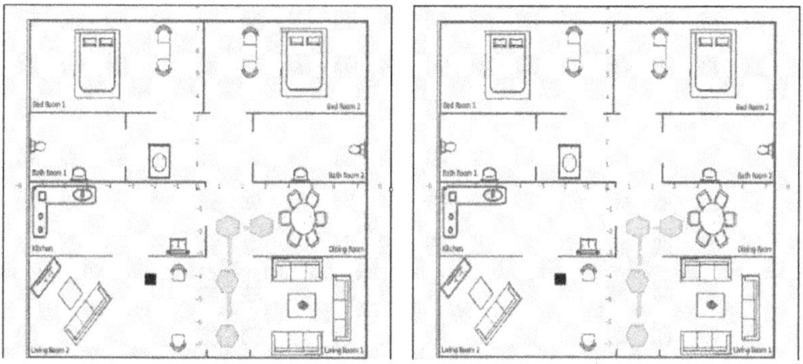

Figure 4.2: Sensor counter Home to D3 - Teaching phase (left) & Execution phase(right) with path (green line)

The system is also implemented in a real-time wheeled mobile robot platform Amrita Autonomous Mobile Robot (AAMoR) shown in Figure 4.5. The robot is trained to navigate an S-shaped trajectory and executes the path accurately in the execution phase. The results are shown in Figure 4.6. As depicted, the paths in both phases are identical, proving that the robot can learn a path and execute it autonomously.

Based on the test results of the proposed counter-based technique, it is evident that a robot can

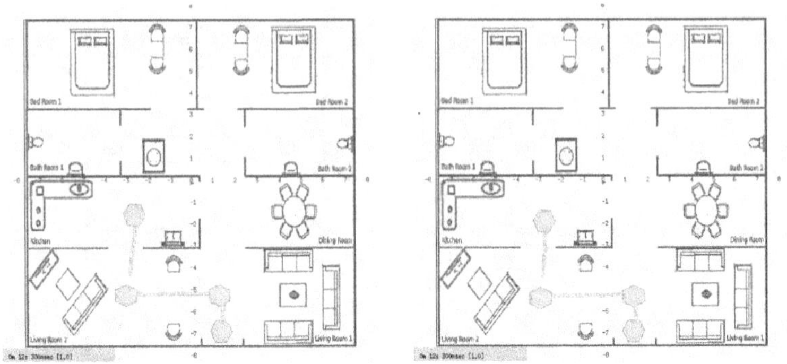

Figure 4.3: Sensor counter Home to D4 - Teaching phase (left) & Execution phase(right) with path (green line)

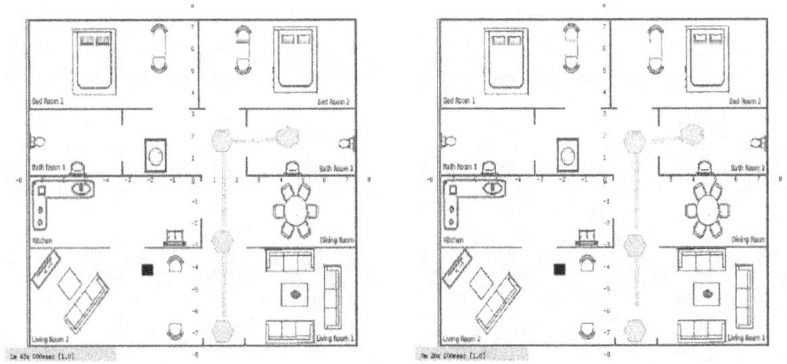

Figure 4.4: Sensor counter Home to D5 - Teaching phase (left) & Execution phase(right) with path (green line)

learn user-trained paths only using onboard sensors. The preliminary proof of concept test was conducted for only one path in the simulator and real-time robot, and the robot could navigate the trained path successfully. The same technique can be extended to train for multiple paths in an environment. There are two drawbacks identified in this technique. First, using a software counter to record the movement will not consider the absolute movement of the wheels while incrementing the counter and is prone to errors like the false increment of the counter. Thus, they do not represent the distance moved accurately. Second, in a wider service environment, the configuration of two different places can be the same; hence, just using the sensor value won't be

Table 4.1: Sensor Counter based technique training accuracy

Path	Training Destination Location	Execution Destination Location	Absolute pose Error
Home→D3	[2.85 -1.85 -92]	[2.95 -1.8 -96]	[0.1 0.05 4]
Home→D4	[-1.8 -3.1 -3]	[-1.9 -3.2 -2]	[0.1 0.1 1]
Home→D5	[3.7 1.5 -88]	[3.5 1.6 -91]	[0.2 0.1 3]

Figure 4.5: Amrita Autonomous Mobile Robot (AAMoR) mobile robot platform

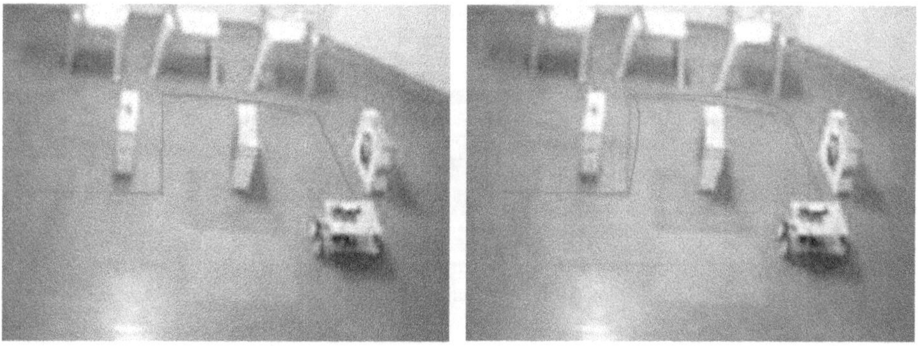

Figure 4.6: Sensor counter S-trajectory Teaching phase (left) & Execution phase (right) with path (red line)

enough to verify the robot's location in a path.

4.1.2 Encoder Based Learning Technique

The encoder-based technique is tested using the Player/Stage simulator. The world and robot are similar to the one used to test the sensor counter method in Section 4.1.1. In the teaching phase, a path is taught, and in the execution phase, the robot has to execute the path autonomously. The robot has been taught three different paths from home position to D3, D4, and D5. Figure 4.7, Figure 4.8, and Figure 4.9 show the teaching and execution phases of the three paths trained. Table 4.2 tabulates all three paths of teaching and execution of final positions and their corresponding errors. The difference in the pose in all the scenarios is very small and negligible. The proposed LfD technique to learn multiple paths using only onboard sensors was successfully implemented and tested.

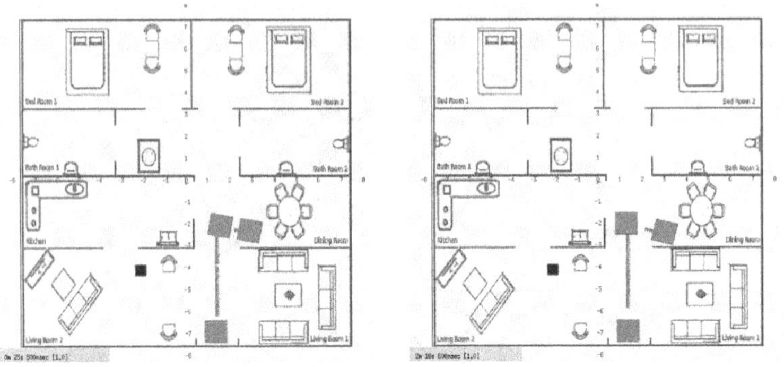

Figure 4.7: EBL Home to D3 - Teaching phase (left) & Execution phase(right) with path (red line)

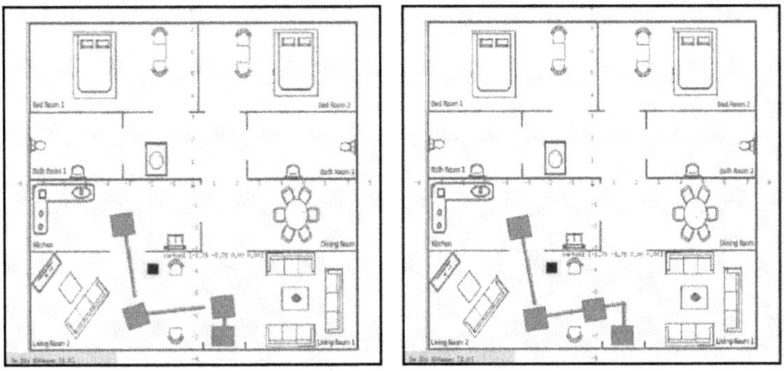

Figure 4.8: EBL Home to D4 - Teaching phase (left) & Execution phase (right) with path (red line)

4.1.3 Enhanced Encoder Based Learning Technique

The EEBL technique is implemented and tested using the player/Stage robotic simulator in the house environment shown in Figure 4.1. In the teaching phase, the robot is trained to reach all eight destinations sequentially, and the resultant path is shown in Figure 4.10. While the path is learned, the robot also identifies reference points. In the execution phase, the robot could navigate to any path autonomously from the home position successfully. The test result is the same as EBL technique as the path matrix generation process is similar except for reference index addition. The key factor to be tested is the ability to auto-correct position and orientation errors.

Table 4.2: EBL technique training accuracy

Path	Training Destination Location	Execution Destination Location	Absolute Pose Error
Home→D3	[3.0 -2.15 -100]	[2.95 -2.25 -106]	[0.05 0.1 6]
Home→D4	[-3.1 -1.9 -5]	[-3.2 -1.95 -4]	[0.1 0.05 1]
Home→D5	[3.6 1.5 -94]	[3.7 1.6 -95]	[0.1 0.1 1]

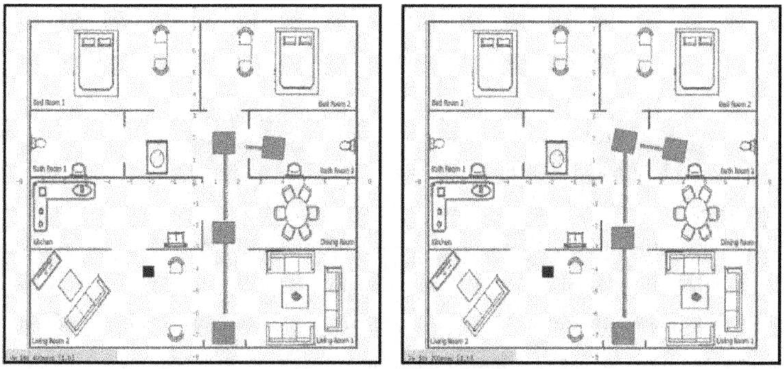

Figure 4.9: EBL Home to D5- Teaching phase (left) & Execution phase (right) with path (red line)

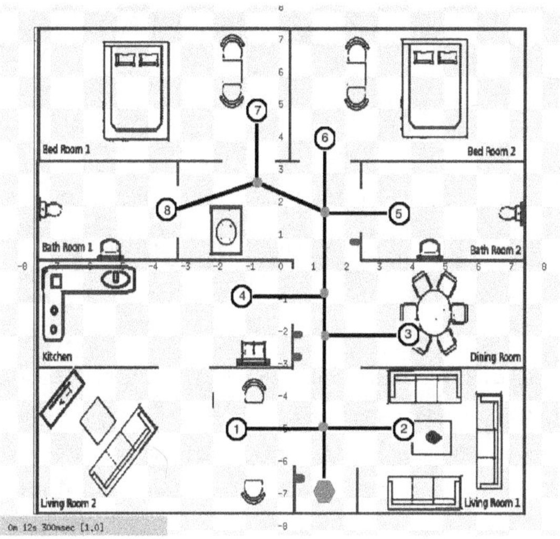

Figure 4.10: House layout with paths learned (black lines), intersections (red circles), and auto-identified reference points (blue circles)

Position and Orientation Error Correction

The most common errors in a mobile robot navigation system are position errors and orientation errors. Orientation error is corrected by storing the pose value of the robot along with path variables in the teaching phase. Position errors can be corrected by identifying reference points along the paths. These references are used in the autonomous phase to correct position errors along the x-axis of the robot's coordinate. A closed-loop PI control system is used to correct the errors. A closed-loop control system automatically regulates a system to maintain a desired state or set point without human interaction. This system auro-corrects the robot's position and orientation to reach the desired position and orientation.

A test was conducted to prove the efficiency of the error correction control system and the repeatability of the proposed system. To test the error correction algorithm efficiency, a Gaussian error is added in the forward movement of the robot (to add a deviation in the heading direction) and the wheel encoder of the robot to introduce an error in the turning angle. Repeatability was verified by running a robot to and fro a fixed source and destination several times and calculating the relative average error of its position. A shadow robot is added in the simulation to run with the actual robot, but without going through any error correction algorithm. This shows the effectiveness of the proposed error correction algorithm. Equation 4.1 shows the robot position representation in the simulation environment. Two paths, Home to D3 and D2 to D5, are taught and tested for error correction. All the destination numbers are prefixed with the letter "D".

$$Robot\_Position = (X, Y, A) \qquad (4.1)$$

where,

X - Robot's X-Coordinate in meters

Y - Robot's Y-Coordinate in meters

A - Robot's Pose Angle in radiance

Case1: Home to D3 Consider the house layout with the source as home with the robot's position as (1.4,-6.8, 0) and destination as D3 with the robot's position as (2.72, -2.02,-1.57). This path was traversed 100 times back and forth continuously with a Gaussian error of μ=0 and σ=1, which was chosen by the trial and error method. Essentially, the robot reaches home and D3 50 times each. Position data for all the trials are collected. Reference position (recorded in the training phase), actual position with error correction, and actual position without error correction are plotted. Figure 4.11. shows the robot's home position plot. Figure 4.12 shows the robot's D3 position plot.

Case2: D2 to D5 In this case, a path from D2 to D5 is traversed 100 times. The robot's position in D2 is (2.64,1.88, -1.466), and the robot's position in D5 is (0.15, -5.11,1.50). This path is mapped by merging two paths from home→D2 and home→D5. This path is traversed back and forth 100 times with Gaussian error of μ=0 and σ=1. The robot's position plot in D2 and D5 after 100 runs are shown in Figure 4.13 and Figure 4.14, respectively.

Case3: Home to D3 with gauss error(0, 2) To validate the efficiency of the error correction algorithm, the same path as case 1 is executed with a different Gaussian error parameter. The robot's position in Home is (1.4,-6.8, 0), and the robot's position in D3 is (2.5, -1.425,-1.57). The

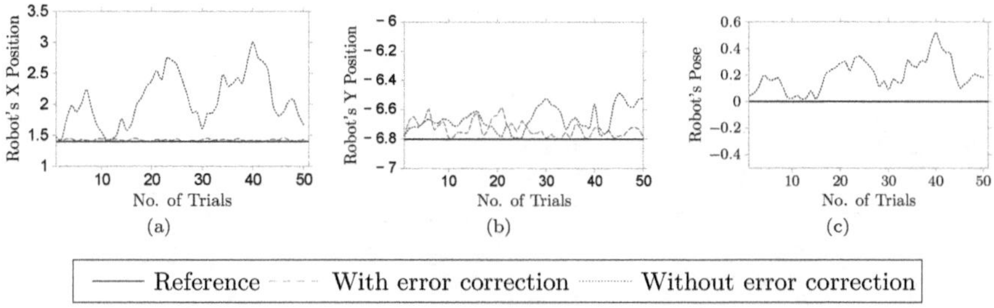

Figure 4.11: Case1: Robot's Home position with Gaussian error (0,1)

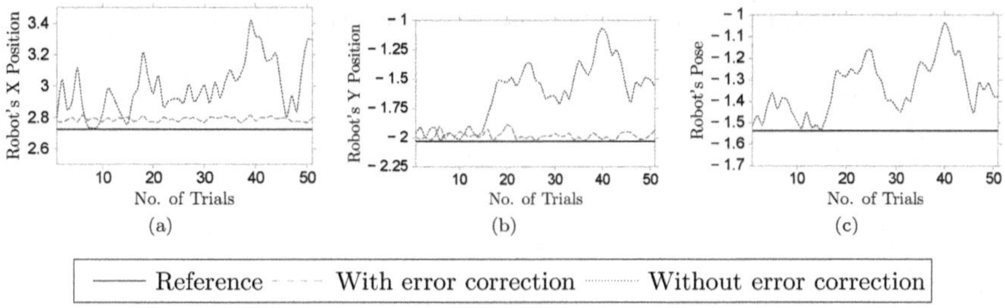

Figure 4.12: Case1: Robot's D3 position with Gaussian error (0,1)

corresponding plot is shown in Figure 4.15 and Figure 4.16.

The average position error for 100 runs is calculated for all the cases. Equation 4.2 calculates the error based on the reference position. Table 4.3 shows the average position error of the cases discussed. The X position and Pose average with error correction is much less than without error correction average. Even though the main objective of our error correction algorithm is to reduce X position and pose angle errors, to an extent, it indirectly corrects Y position errors, too. This can be observed in the case1:D3 and case3:D3 scenarios. The average positional error is shown

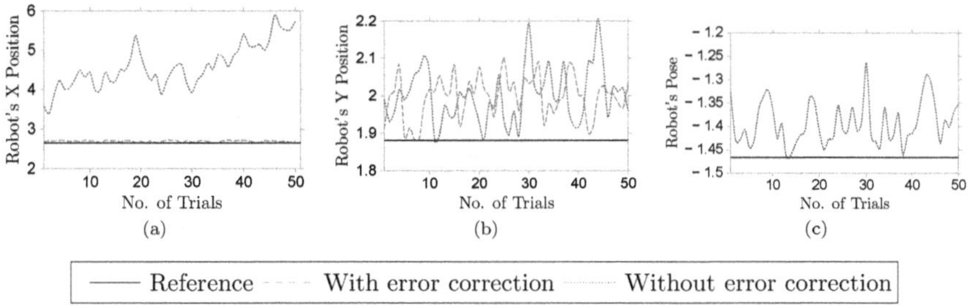

Figure 4.13: Case2: Robot's D2 position with Gaussian error (0,1)

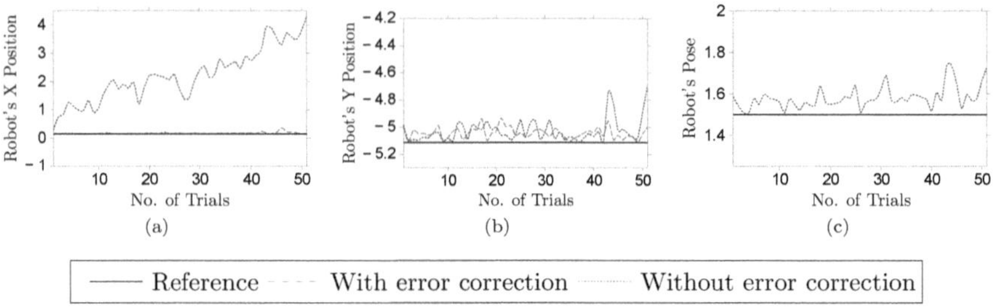

Figure 4.14: Case2: Robot's D5 position with Gaussian error (0,1)

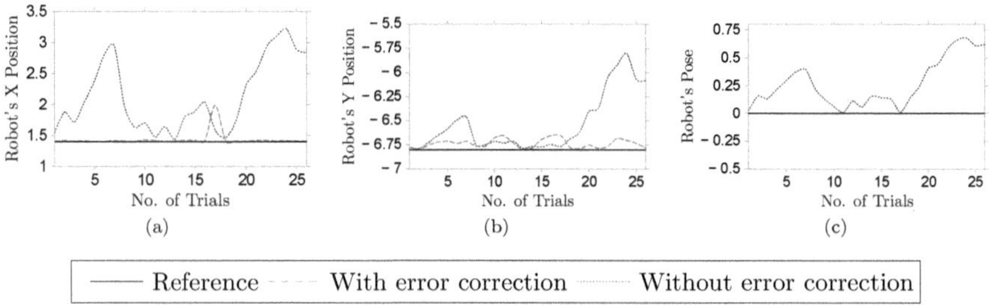

Figure 4.15: Case3: Robot's Home position with Gaussian error (0,2)

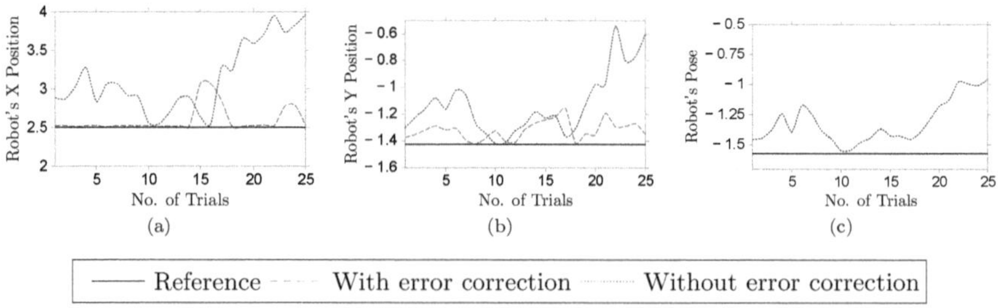

Figure 4.16: Case3: Robot's D3 position with Gaussian error (0,2)

Table 4.3: Average error

Test Case	Location	With error correction			Without error correction		
		X	Y	A	X	Y	A
Case1	Home (1.4,-6.8, 0)	0.02	0.069	0	0.67	0.13	0.19
	D3 (2.72, -2.02,-1.57)	0.06	0.04	0	0.26	0.40	0.19
Case2	D2 (2.64,1.88, -1.466)	0.04	0.11	0	1.94	0.11	0.07
	D5 (0.15, -5.11,1.50)	0.02	0.09	0	2.05	0.08	0.08
Case3	Home (1.4,-6.8, 0)	0.03	0.25	0	0.73	0.26	0.26
	D3(2.5, -1.425,-1.57)	0.09	0.1	0	0.63	0.29	0.25

in Equation 4.3. This is calculated using a column-wise average from Table 4.3. From this, it is evident that the proposed error correction system maintains the robot's position close to the reference position.

$$(Avg\_error)_i = \frac{|Reference_i - Actual_i|}{\text{Total no. of runs}} \tag{4.2}$$

where,

i = X or Y or A

$Reference_i$ = Reference value

$Actual_i$ = Actual value

$$Avg\_Pos\_Error\_With\_Correction = (0.04, 0.11, 0)$$
$$Avg\_Pos\_Error\_Without\_Correction = (1.05, 0.21, 0.17) \tag{4.3}$$

4.2 Path Mapping and Planning

The proposed algorithms are implemented in the player/stage 2D robotic simulation platform. The system is tested in two different environment layouts, a house and a hospital, as shown in Figure 4.17. The House layout is a smaller environment with 8 destinations, and the hospital is comparatively a bigger and more complex and bigger environment with 28 destinations.

Initially, the robot is assigned a home position in the environment, and sequential training to navigate all the destinations is carried out using the Enhanced Encoder Based Learning (EEBL)

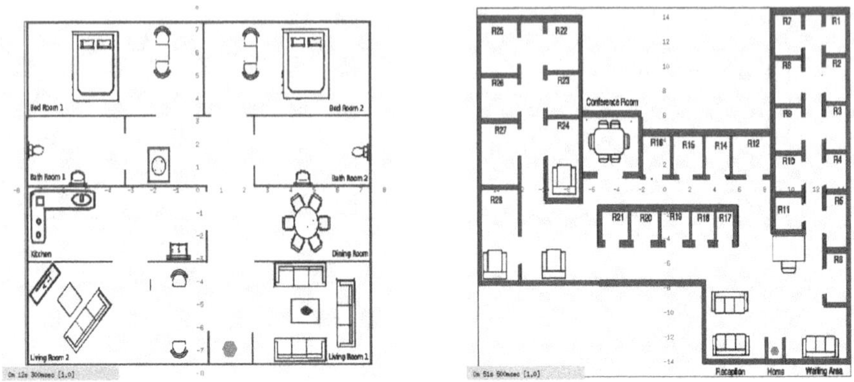

Figure 4.17: Stage environment with robot(hexagon shaped red object) in home position, house layout(left) and hospital layout(right)

algorithm. While the paths are being learned, the robot also identifies Virtual Land Mark (VLM) and reference points in the environment. After the training phase, the TBP algorithm generates a tree with the home as the root node, VLM's as intermediate nodes, and all the destinations as child nodes.

Figure 4.18 depicts the paths in the house layout with VLM's and reference points, and Figure 4.19 shows the tree generated for the paths learned with 5 VLM's and 8 destinations. Figure 4.20 shows the paths learned in the hospital layout with VLM's and reference points. Figure 4.21 shows the tree formed in the hospital layout with 23 VLM's and 28 destinations. Compared to the house layout, this layout is 3.5 times bigger than the number of destinations and nodes in the tree. This proves that the algorithm is scalable. Given a destination, a path can be planned using the tree, and the same can be navigated to reach the destination. For example, consider the robot's current position is in D1 of the hospital layout, and the given destination is D21. In this case, the robot plans a path by traversing the tree and finding the shortest possible route from source to destination, which is D1→V8→V7→V6→V5→V4→V2→V1→V3→V9→V10→V11→V12→V13→D21.

The proposed system is compared with breadth-first search (BFS), Dijkstra's, greedy best-first search (BBF), A*, and rapidly exploring random trees(RRT) algorithm. To compare with the proposed algorithm, a Java-based path planning simulator is used (Dhanjal, 2016). A replica (with the same number of cells) of the hospital layout is created in the Java simulator, and the resultant path is compared. The exploratory path planning algorithm searches the environment to find an efficient path to the given environment. In the simulator, each cell's search space is denoted by yellow and black squares. The yellow squares represent all the searched cells, and the black cells represent the search space's boundary.

In the hospital layout, three paths considered are Home→D21, Home→D28, and D7→D28. The path mapped by the proposed algorithm is given in Equation 4.4. Figure 4.22, Figure 4.23, and

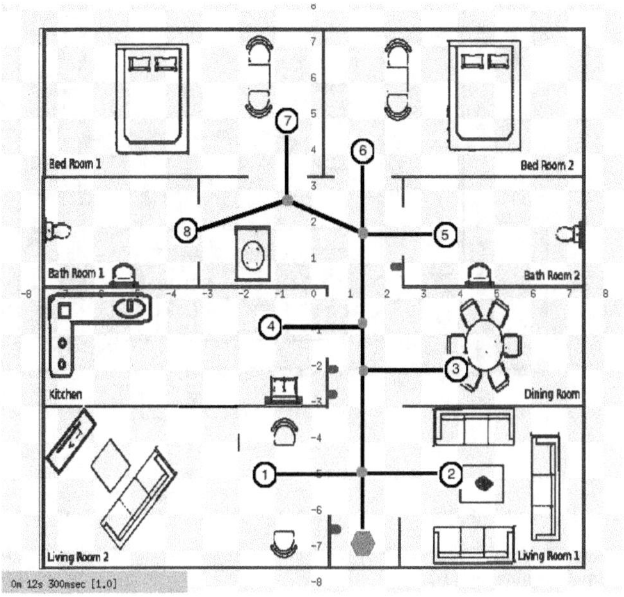

Figure 4.18: House layout with paths learned (black lines), identified VLM's (red circles) and reference points(blue circles)

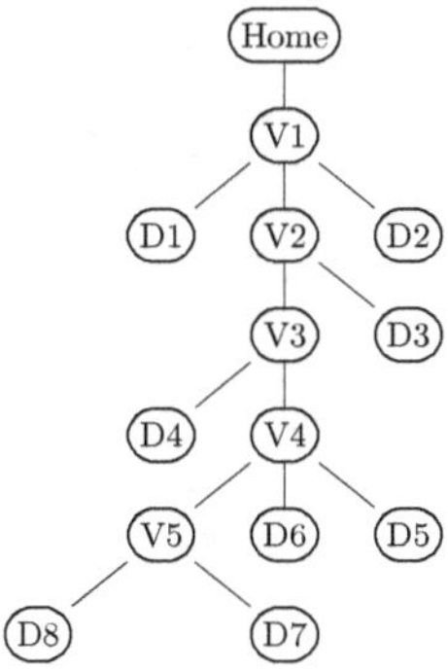

Figure 4.19: Tree formed based on house layout paths

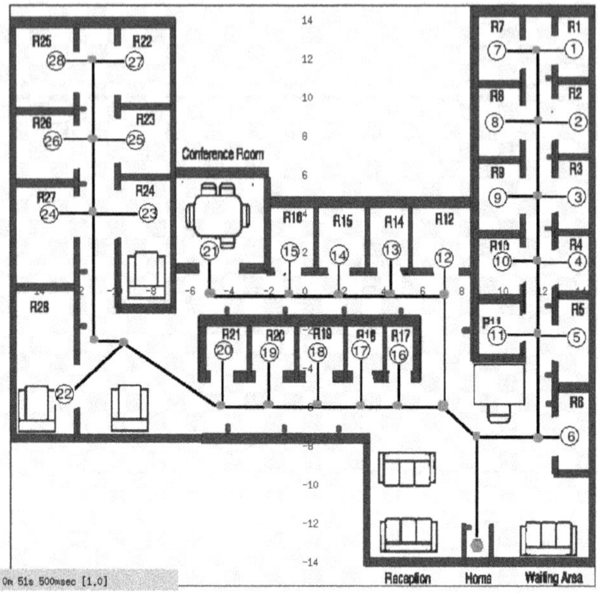

Figure 4.20: Hospital layout with paths learned (black lines), identified VLM's (red circles) and reference points(blue circles)

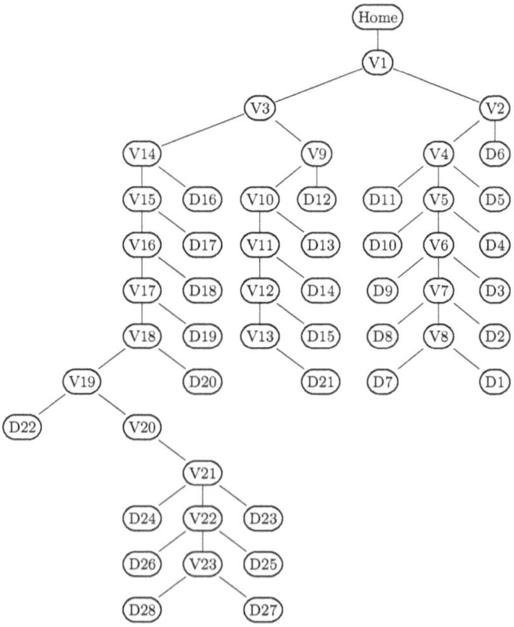

Figure 4.21: Tree formed based on hospital layout paths

Figure 4.24 show the path generated by the proposed and other algorithms for three different paths. Similarly, in the house layout, three paths considered are Home→D6, D7→D3, and Home→D7. Figure 4.25, Figure 4.26, and Figure 4.27 show the path generated by the proposed and other algorithms for three different paths.

$$Path1 : Home \rightarrow V1 \rightarrow V3 \rightarrow V9 \rightarrow through \rightarrow V13 \rightarrow D21 \tag{4.4a}$$

$$Path1 : Home \rightarrow V1 \rightarrow V3 \rightarrow V14 \rightarrow through \rightarrow V22 \rightarrow D28 \tag{4.4b}$$

$$\begin{aligned} Path2 : D7 &\rightarrow V8 \rightarrow V7 \rightarrow V6 \rightarrow V5 \rightarrow V4 \rightarrow V2 \rightarrow V1 \\ &\rightarrow V3 \rightarrow V14 \rightarrow through \rightarrow V22 \rightarrow D28 \end{aligned} \tag{4.4c}$$

A quantitative comparison of the paths generated in both the layouts with other algorithms based on path length and visited cells is depicted in Table 4.4 and Table 4.5. The path length is the number of cells the robot's path is traced to reach the destination. Visited cells are the number of cells the path planning algorithm visited to find the optimal path to reach the destination. The difference percentage error calculated for all the algorithms against the proposed algorithm is also tabulated. The average difference percentage for all three paths in the hospital layout is 10.6% for path length and 115.7% for visited cells. Similarly, in house layout, it is calculated to be 3.4% for path length and 125.3% for visited cells. This proves that the path length mapped by the proposed algorithm is on par with the other algorithms, whereas the visited cell count is very small compared to other algorithms. This is mainly due to the learning in the LfD stage. This considerably reduces the path planning time in the proposed algorithm. Hence, the proposed algorithm is faster with efficient path mapping and planning.

4.2.1 State Based Obstacle Avoidance Algorithm

The obstacle avoidance algorithm has been tested using various cases using the Player/Stage simulator. The algorithm can help the robot to overcome different kinds of obstacles. Figure 4.28 & Figure 4.29 show two scenarios where the robot overcomes two obstacles and returns to the original path.

Figure 4.28 depicts the robot avoiding a U-shaped obstacle avoidance, and the states it goes through are D-A-DH-AH-A-C. The black line shows the robot's path, avoiding obstacles and converging back to its path. Figure 4.29, the robots avoid a stepwise obstacle, and the states are D-A-D-A-D-

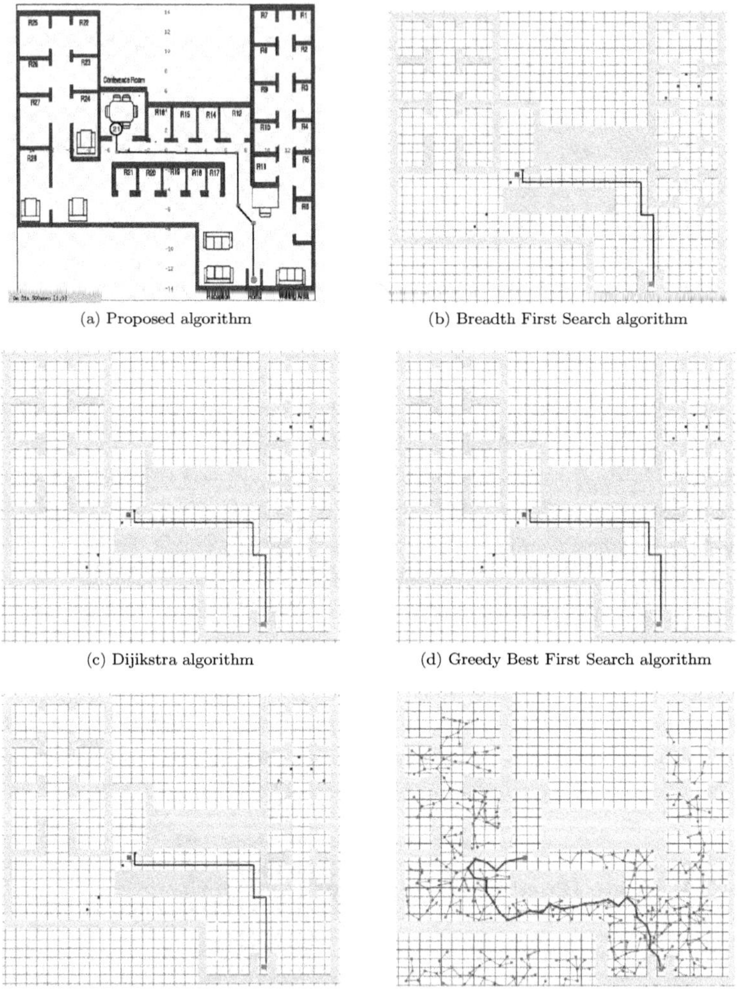

(a) Proposed algorithm

(b) Breadth First Search algorithm

(c) Dijikstra algorithm

(d) Greedy Best First Search algorithm

(e) A* algorithm

(f) RRT algorithm

Figure 4.22: Path 1: Hospital layout from Home to D21

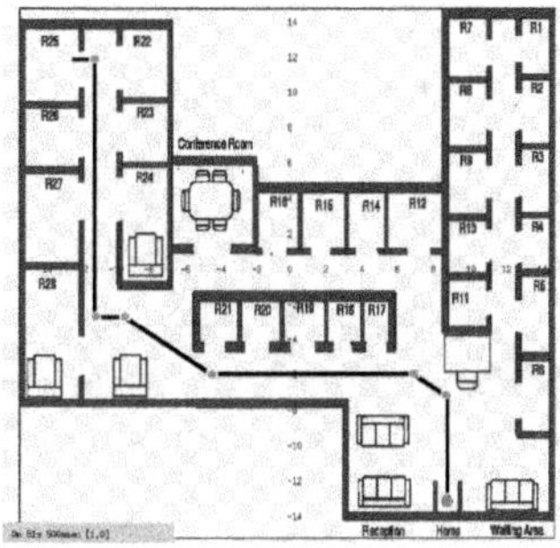

(a) Proposed algorithm

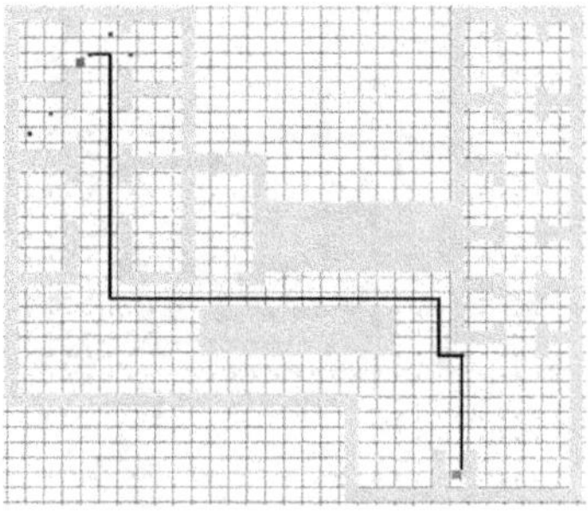
(b) Breadth First Search algorithm

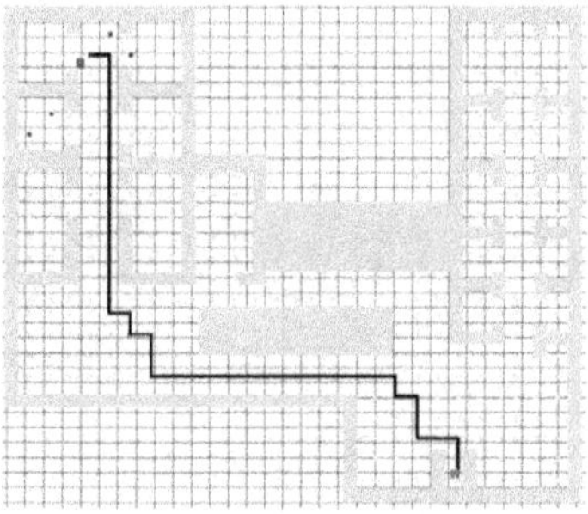
(c) Dijikstra algorithm

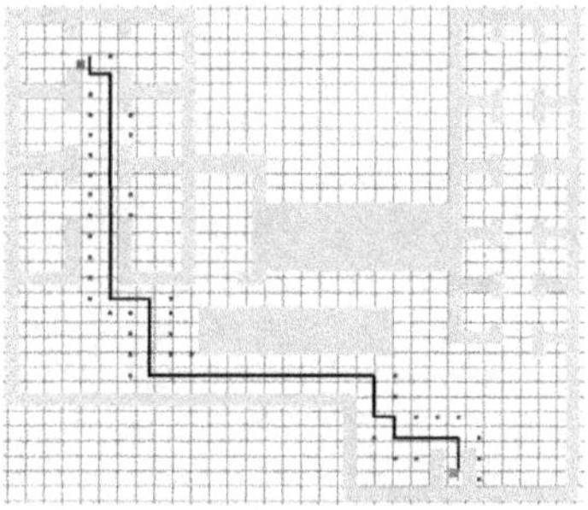
(d) Greedy Best First Search algorithm

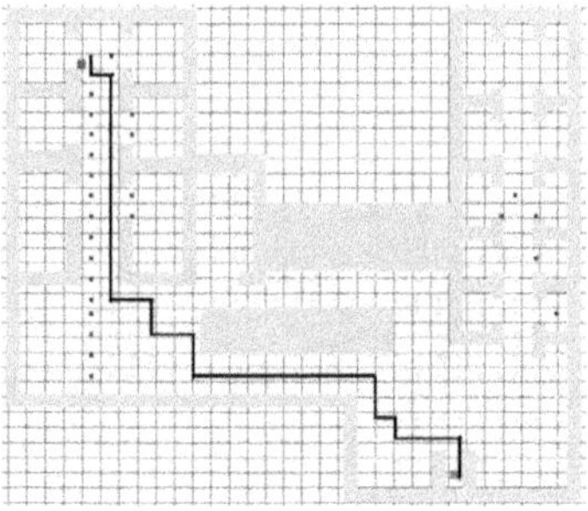
(e) A* algorithm

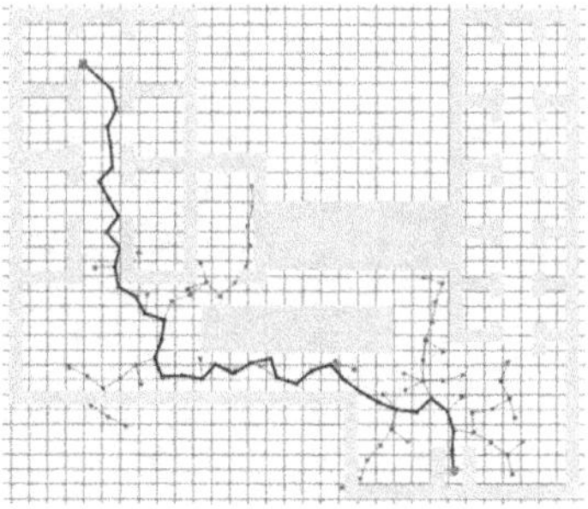
(f) RRT algorithm

Figure 4.23: Path 2: Hospital layout from Home to D28

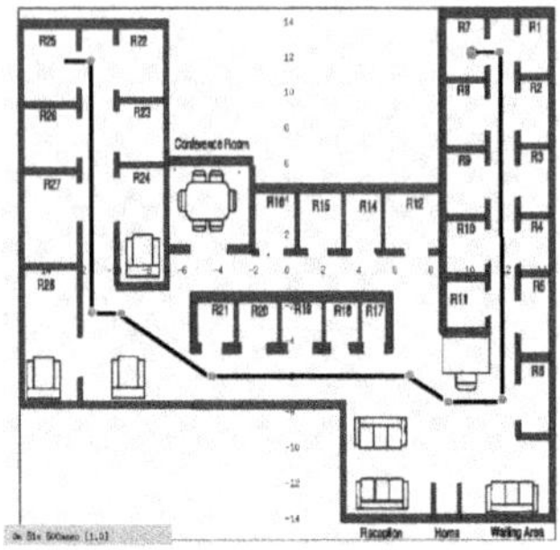

(a) Proposed algorithm

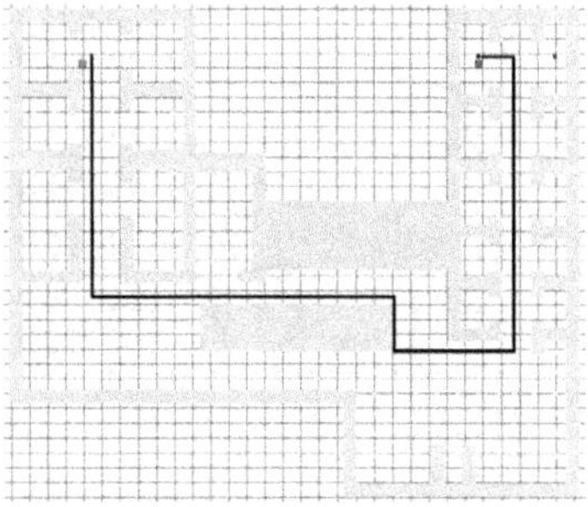

(b) Breadth First Search algorithm

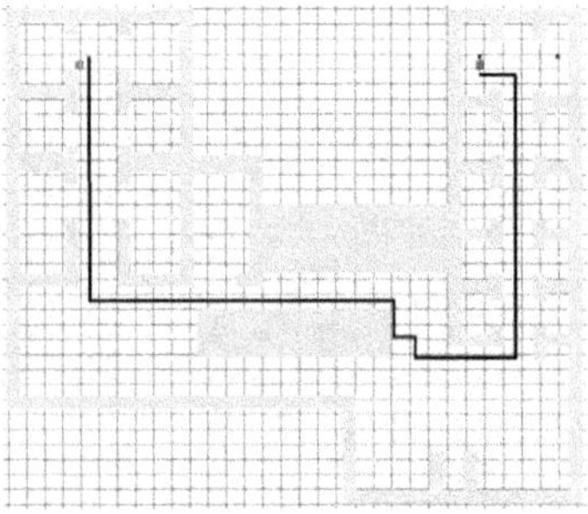

(c) Dijikstra algorithm

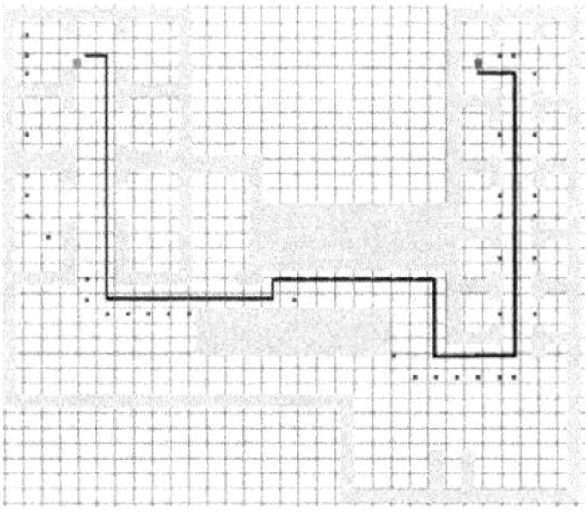

(d) Greedy Best First Search algorithm

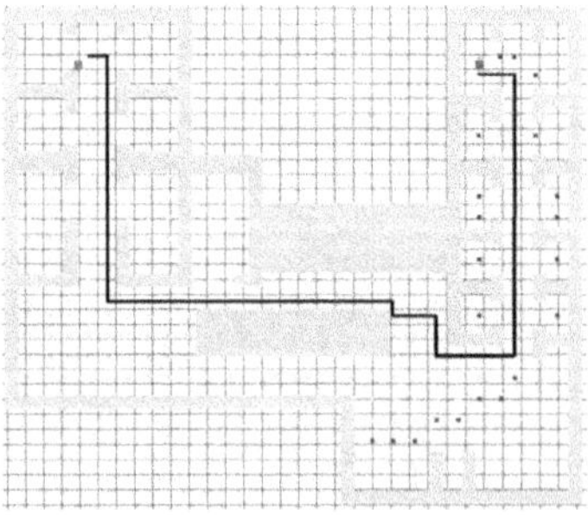

(e) A* algorithm

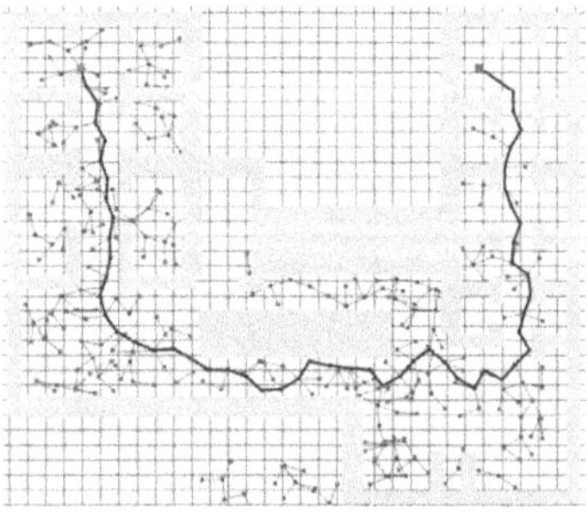

(f) RRT algorithm

Figure 4.24: Path 3: Hospital layout from D7 to D28

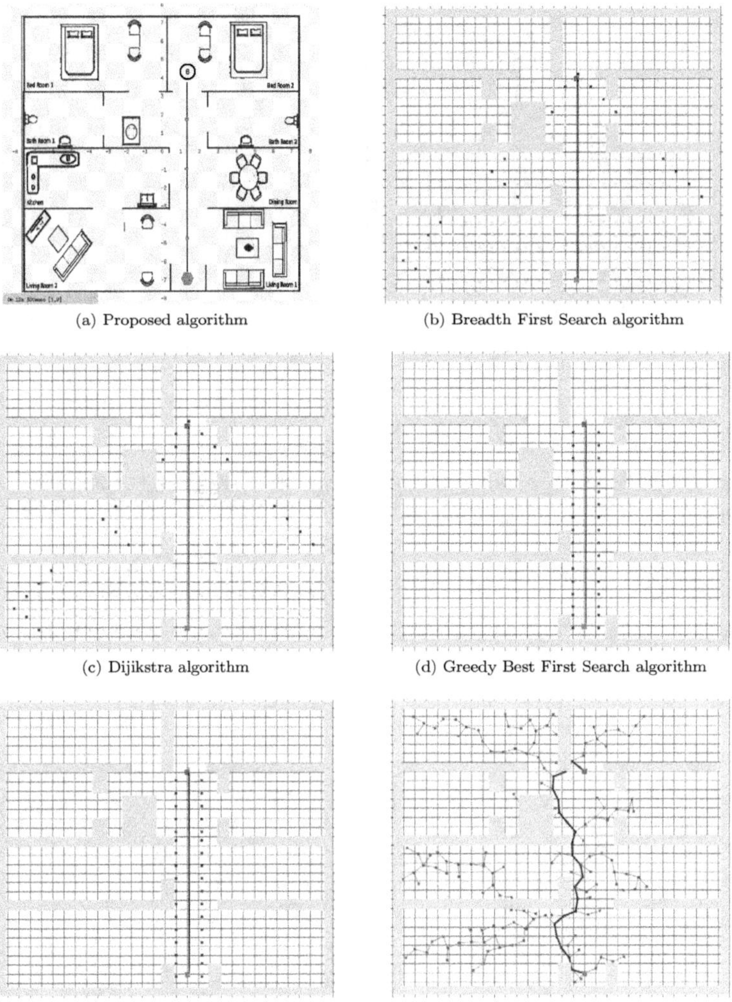

(a) Proposed algorithm

(b) Breadth First Search algorithm

(c) Dijikstra algorithm

(d) Greedy Best First Search algorithm

(e) A* algorithm

(f) RRT algorithm

Figure 4.25: Path 1: House layout from Home to D6

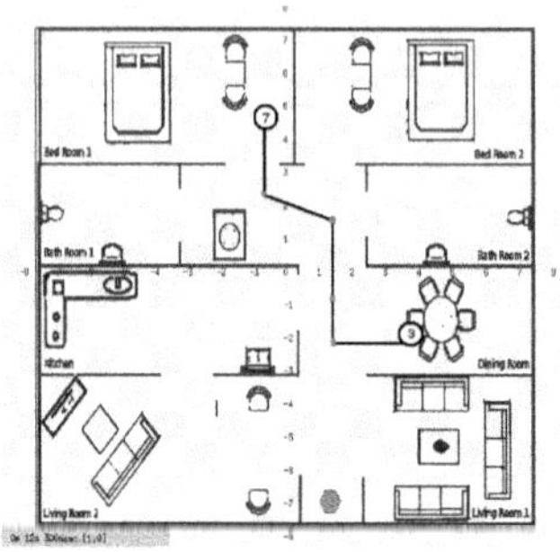

(a) Proposed algorithm

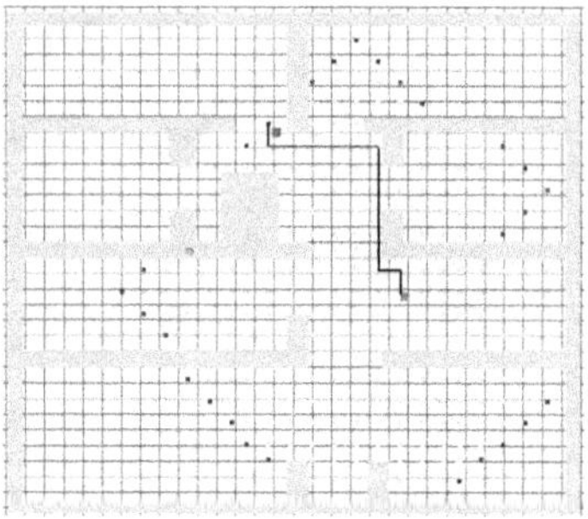

(b) Breadth First Search algorithm

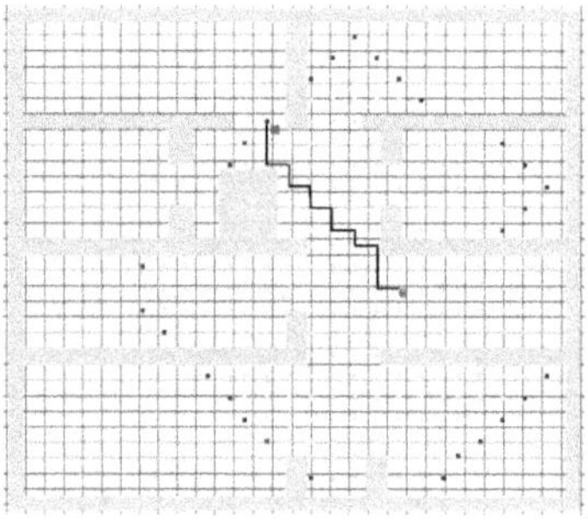

(c) Dijikstra algorithm

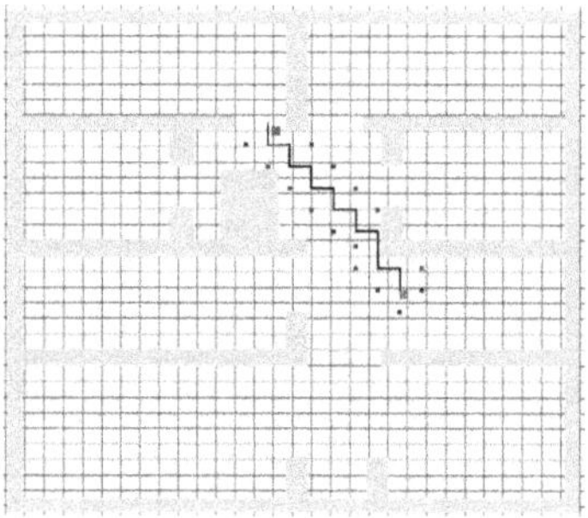

(d) Greedy Best First Search algorithm

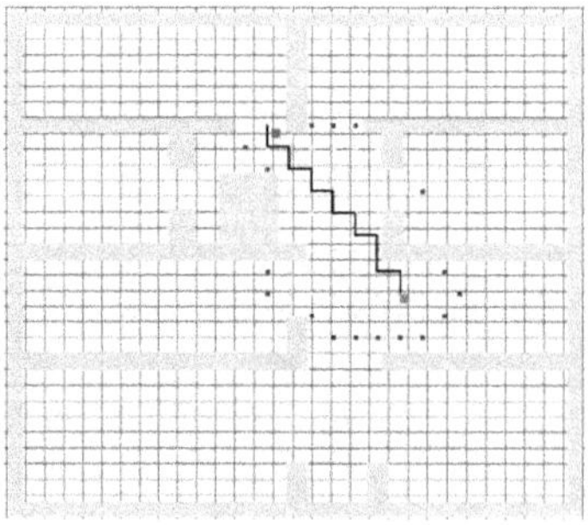

(e) A* algorithm

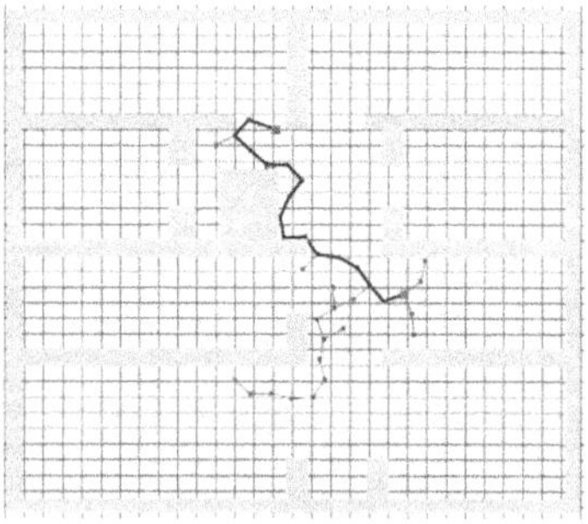

(f) RRT algorithm

Figure 4.26: Path 2: House layout from D7 to D3

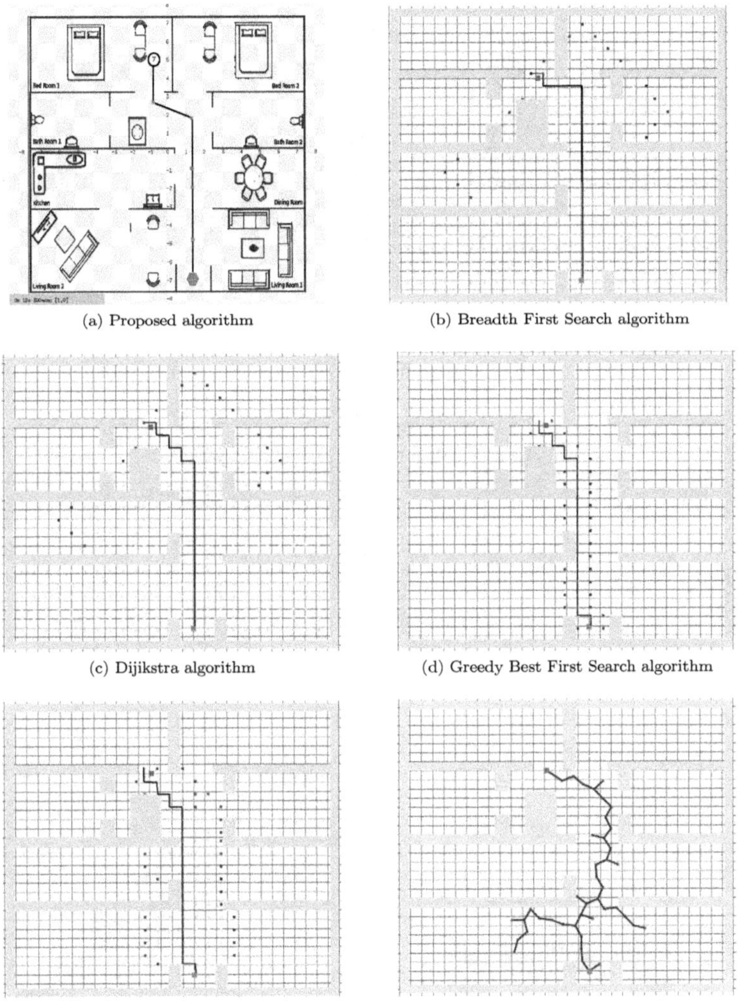

(a) Proposed algorithm

(b) Breadth First Search algorithm

(c) Dijikstra algorithm

(d) Greedy Best First Search algorithm

(e) A* algorithm

(f) RRT algorithm

Figure 4.27: Path 3: House layout from Home to D7

Table 4.4: Hospital layout quantitative path analysis

Paths	Algorithm	Length (cell count)	Visited Cells	% Difference Length	% Difference Visited Cells
Hospital Layout Path 1	Proposed	25	25	-	-
	BFS	25	143	0	140.4
	Dijkstra	25	143	0	140.4
	GBF	25	44	0	55.07
	A*	25	86	0	109.9
	RRT	33	323	27.5	123.4
	Path1 Average % Difference			**5.5**	**123.4**
Hospital Layout Path 2	Proposed	41	41	-	-
	BFS	48	274	15.7	147.9
	Dijkstra	47	274	13.6	147.9
	GBF	48	76	15.7	59.8
	A*	25	188	48.4	128.3
	RRT	42	101	2.4	85.5
	Path2 Average % Difference			**19.1**	**113.7**
Hospital Layout Path 3	Proposed	59	59	-	-
	BFS	63	205	6.5	110.6
	Dijkstra	62	205	4.9	110.6
	GBF	64	165	8.1	94.6
	A*	62	236	4.9	120
	RRT	66	215	11.2	113.8
	Path3 Average % Difference			**7.1**	**109.9**
Hospital Paths Average % Difference				**10.6**	**115.7**

Table 4.5: House layout quantitative path analysis

Paths	Algorithm	Length (cell count)	Visited Cells	% Difference Length	% Difference Visited Cells
House Layout Path 1	Proposed	23	23	-	-
	BFS	23	231	0	163.7
	Dijkstra	23	231	0	163.7
	GBF	23	51	0	75.67
	A*	23	51	0	75.67
	RRT	27	204	16	159.47
	Path1 Average % Difference			**3.2**	**127.67**
House Layout Path 2	Proposed	16	16	-	-
	BFS	17	242	6.0	175.1
	Dijkstra	17	233	6.0	174.2
	GBF	17	29	6.0	57.7
	A*	17	68	6.0	123.8
	RRT	18	60	11.7	115.7
	Path2 Average % Difference			**7.2**	**129.3**
House Layout Path 3	Proposed	26	26	-	-
	BFS	26	311	0	169.1
	Dijkstra	26	311	0	169.1
	GBF	26	51	0	64.9
	A*	26	105	0	120.6
	RRT	26	55	0	71.6
	Path3 Average % Difference			**0**	**125.3**
House Paths Average % Difference				**3.4**	**125.3**

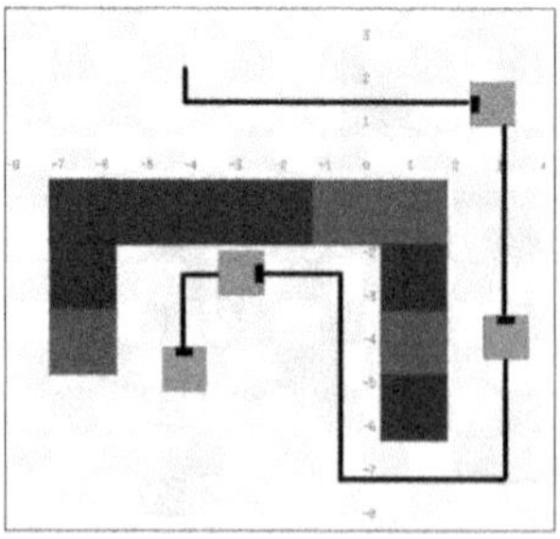

Figure 4.28: Obstacle avoidance scenario 1

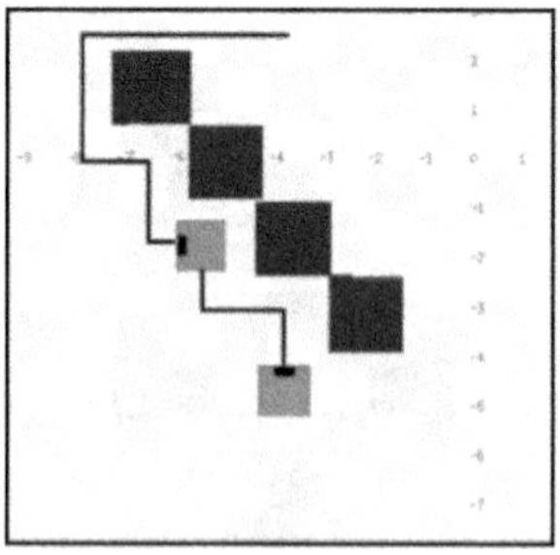

Figure 4.29: Obstacle avoidance scenario 2

A-C. This scenario depicts the piecewise linear movement for a step-wise obstacle. In both cases, the robot avoids the obstacle and returns to its original path.

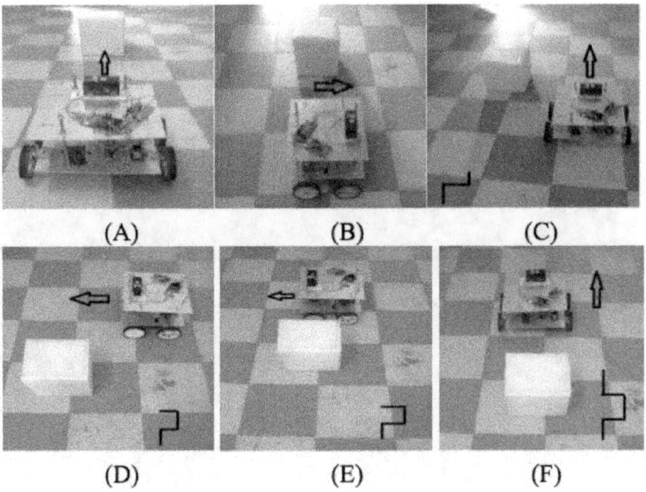

Figure 4.30: AAMoR-I avoiding a simple square-shaped obstacle

The algorithm is implemented and tested in AAMoR mobile robot platform. A box-shaped obstacle avoidance states are depicted in Figures 4.30A to F. In A, the robot detects the obstacle, then in B, it goes to diverge state, and in C it is in avoid state. Finally, D to F is in the converge state. Similarly, a "U" shaped obstacle avoidance states are depicted in Figures 4.31 A to F. Here, A shows obstacle detection, B in the diverged state, C in the diverging horizontal state, D in the avoid horizontal state, E in the avoid state and F in converge state. In both cases, the robot avoids the obstacle and reverts to its original path.

Different obstacle patterns are introduced in the mobile robot's path to test the algorithm. It includes a simple rectangular box structure, a Step obstacle, and an inverted 'U' shape obstacle as shown in Figure 4.32(A - E). The proposed algorithm is implemented & tested successfully. The mobile robot avoids objects in its path and reverts to its original path.

The proposed path mapping and planning system are scalable and efficient based on the test results, and comparison results with a few state-of-the-art path planners discussed. The algorithm is tested and proved to avoid different obstacle configurations efficiently. The system works well as long as the environment configuration is the same as in the learning stage. If there is a change in the environmental configurations, the changes must be incorporated to obtain efficient paths to traverse.

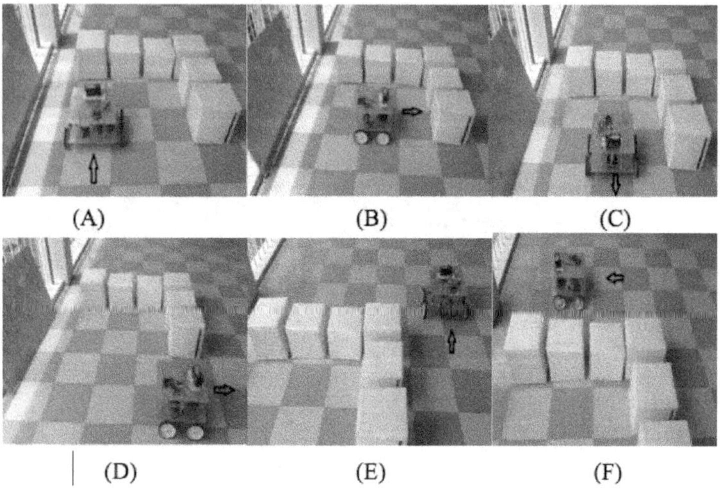

Figure 4.31: AAMoR-I avoiding inverted U-shaped obstacle

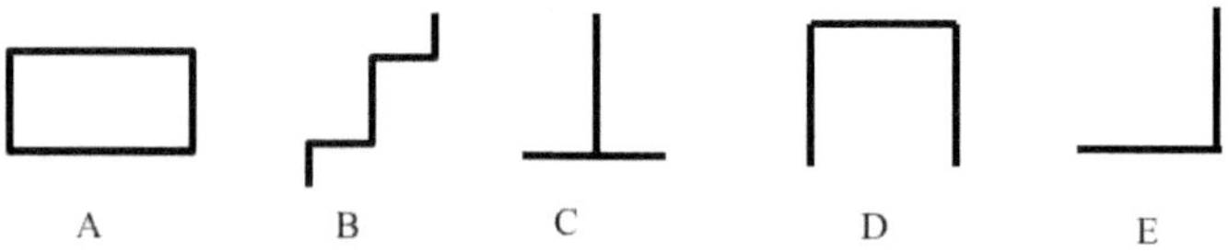

Figure 4.32: Different types of obstacle shapes tested

4.3 Lifelong Learning

The RL path planner is proposed to be implemented in two stages: pre-training and lifelong learning. The pre-training stage trains the RL agent to plan efficient paths based on the topological map. The lifelong learning stage uses the β-decay Transfer Learning Algorithm to keep the RL agent in continuous learning mode to update its training based on environmental changes. A Q-learning-based path planner is implemented and tested with one destination and multi-destinations to test the viability of the proposed RL-based path planner. After the successful implementation of the Q-learning base path planner, a more scalable Deep Q-Network (DQN)-based path planner is implemented and tested. The system is implemented in the ROS2 framework and tested using Turtlebot3 mobile in the Gazebo simulation environment.

4.3.1 Q-Learning Based Path Planner

The proposed path planner is trained and tested in two different service environments. One environment has a house layout with eight destinations, and another has a hospital layout with 28, as shown in Section 4.2 Figure 4.17. A new node named "U" is added for the agent to learn the environment completely. This node represents the unknown state of the agent. So, the topological map used for training is a complete ternary tree with unknown leaf nodes. This enables the agent to stop taking actions transiting to the unknown state. A sample environment and its corresponding topological map with "U" states are depicted in Figure 4.33 and Figure 4.34, respectively. The training hyper-parameters α, γ, and, β are chosen to be 0.9 based on experimental trials. All the parameters are identical for both environments except for the training episodes. More nodes are in the hospital layout, so more training episodes exist.

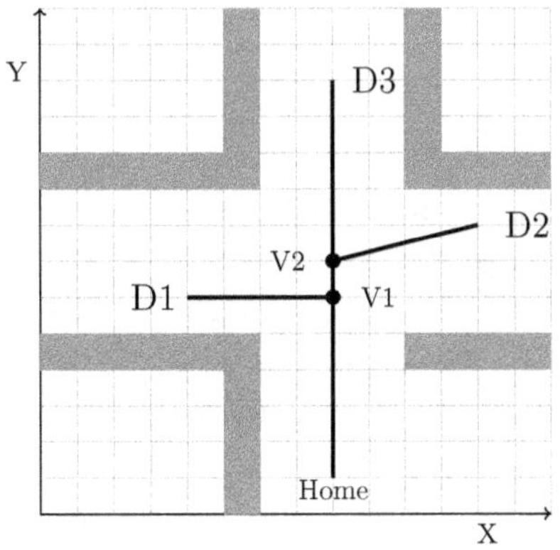

Figure 4.33: Sample environment with paths to multiple destinations

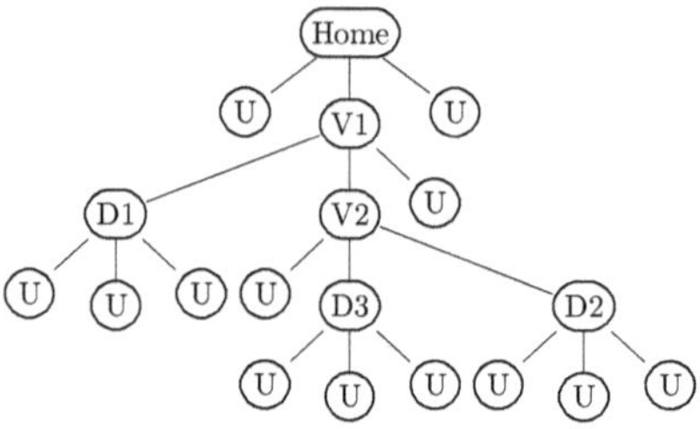

Figure 4.34: Topological map with "U" states

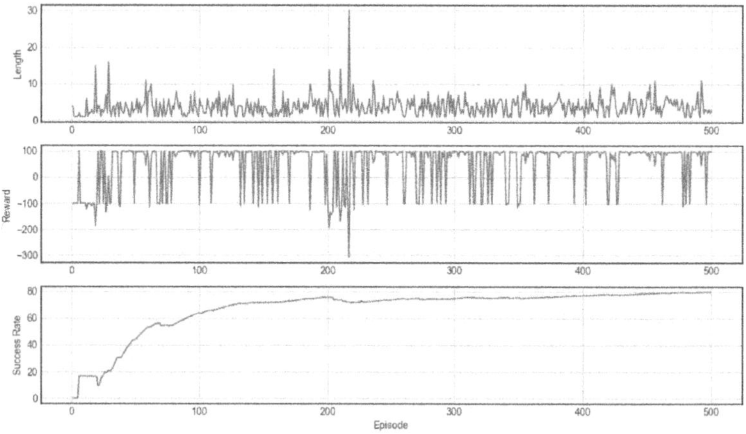

Figure 4.35: House layout destination "D8" training metrics

House Layout Training

Single Goal Training: Initially, the RL agent was trained to reach one of the destinations to fine-tune the hyper-parameters and to identify the minimum training episodes required to converge fully. D8 in the house layout is chosen for training. Figure 4.35 shows the plot of the training metrics for 500 episodes. The training success rate estimates the agent's convergence level over the total training episodes. The training success rate is 79% for 500 episodes of training.

Multi-Goal Training: Generally, RL problems are defined as single-goal problems. The RL agent is trained to converge to attain a single goal with maximum rewards. But for a path planning problem, there can be multiple destinations a robot is required to reach autonomously. A path-planning Q-learning agent can be trained to plan the path for multiple destinations by

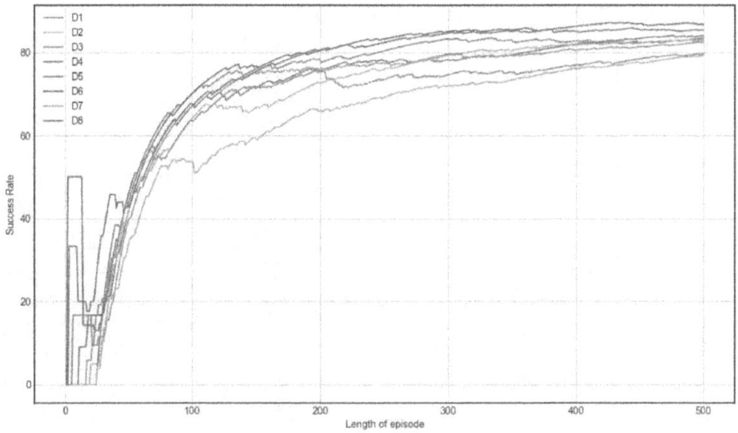

Figure 4.36: House layout success rate

Table 4.6: House layout success count and success rate for all the destinations

Destination Name	Success Episode Count (out of 500)	Success Rate
D1	417	83.39
D2	415	83.00
D3	411	82.36
D4	427	85.39
D5	420	84.00
D6	433	86.60
D7	396	79.20
D8	398	79.60

changing the Q-table and next state table to n dimensions where n is the total number of goals. The training process is similar to single-goal training, where training is carried out for all the destinations sequentially. The success rate plot for all the training episodes is plotted destination-wise in Figure 4.36. Each destination's training success rate and success count are tabulated in Table 4.6. The agent achieved an accuracy of 86.6% average success rate for all the goals.

Hospital Layout Training

A multi-goal Q-learning agent was configured based on the hospital layout environment. The agent's training was carried out sequentially with 3000 episodes per destination. More training episodes exist than in the house layout as more nodes are covered. The training success rate and success count for each destination are shown in Table 4.7. The agent achieved an accuracy of 70.4%

Table 4.7: Hospital layout success count and success rate for all the destinations

Destination Name	Success Episode Count (out of 3000)	Success Rate
D1	2060	68.66
D2	2160	72.00
D3	2230	74.33
D4	2159	71.99
D5	2241	74.70
D6	2259	75.30
D7	2003	66.76
D8	2210	73.66
D9	2214	73.80
D10	2231	74.36
D11	2300	76.66
D12	2201	73.36
D13	2069	68.96
D14	2067	68.92
D15	2031	67.70
D16	2243	74.81
D17	2213	73.79
D18	2202	73.40
D19	2151	71.70
D20	2030	67.66
D21	1782	59.40
D22	2026	67.55
D23	1938	64.60
D24	1992	66.40
D25	1952	65.06
D26	1849	61.63
D27	1577	52.56
D28	1769	59.00

average success rate for all the goals.

Testing

Both agents were tested after successfully training the RL-based path planner agent in two different environments. The testing evaluated all the paths generated given a [Source, Destination] location pair. The source location can be any destination node, and the source node can be any other destination node. Testing is carried out with all the combinations of source and destination as destination nodes. Table 4.8 gives a comparison summary of all the crucial testing and training parameters of both environments. The key takeaway from the table is that even though there is a vast difference in the total number of nodes and destinations for both environments, a Q-learning agent with the same hyper-parameters setting could learn to plan the path efficiently with a short duration of training.

Table 4.8: Multi-goal testing & training parameters

Parameter	House Layout	Hospital Layout
No. of destination	8	28
No. of training episodes/destination	500	3000
Average training time/destination	2.4 secs	48 secs
Average training success rate	86.6%	70.4%
No. of test cases	64	784
Accuracy	98.43%	95.57%

4.3.2 Deep Q-Learning Based Path Planner with Transfer Learning

As shown in Section 4.3.1, two service environments are used to test the system. To test the system's scalability, a benchmark maze environment is used to test the proposed Deep Q-Network (DQN) framework. The performance of the Deep Reinforcement Learning (DRL) framework is measured based on its training and testing success rate. The Transfer Learning (TL) module performance is measured based on three parameters: jumpstart, time-to-threshold, and time-to-converge.

Deep Q-Learning (DQL) Path Planner Pre-Training

The Deep Q-Learning (DQL) planner is pre-trained using the topological map of the environment. The system is trained and tested in two service environments to prove the path planner's scalability and generality.

Selection of Hyper-Parameters: One important step in training any RL agent is to tune the hyper-parameters, which enables the agent to learn the task efficiently. The hyper-parameters are learning rate-α, discount factor-γ, range exploration factor-ϵ, and replay memory batch size. The pace of the agent's learning is based on the learning rate. The learning rate is often chosen between values 0-1. A higher learning rate will result in less learning time but at the cost of sub-optimal learning quality. A lower learning rate will result in optimal learning quality but at the cost of increased learning time. Since the proposed RL agent is trained for a service environment and is carried out offline, learning quality is chosen over learning time. Hence, α is 0.001 to achieve stable and quality learning. The discount factor-γ is used to evaluate the weight of future rewards. γ is chosen at 0.95 to give importance to future rewards while learning.

The other hyperparameters, ϵ range, and batch size are determined by performing single-goal training in the house layout. In the house layout, the RL agent is trained to reach two destinations, D1 and D8, for 500 episodes. Batch size values 32 and 16 were chosen, as the neural network is trained with numerical values. ϵ range is given as (ϵ_{intial} - ϵ_{min}), a range of numbers between an initial value and minimum value with exponential decay factor calculated based on Equation 3.16. ϵ-ranges chosen are (1 - 0.1) and (0.1 - 0.001). The evaluation parameter is the training success

Table 4.9: ϵ-range & batch size success rate

Batch Size	ϵ-range			
	Destination-D1		Destination-D8	
	1-0.1	0.1-0.001	1-0.1	0.1-0.001
32	81.6%	48.4%	75.6%	43%
16	65.5%	47.8%	36.1%	31%

Table 4.10: Hyper-Parameters

Parameter	Values
α	0.001
γ	0.95
ϵ-range	(1-0.1)
Batch Size	32

rate, which indicates the agent's convergence. It is calculated based on the ratio of successful and total training episodes. Table 4.9 shows the training success rates of all the training cases, which vary based on batch size and ϵ range.The test case with batch size 32 and ϵ range of (1 - 0.1) yielded higher success rates (marked in red). Based on this, the final hyper-parameter values are tabulated in Table 4.10.

House Layout Training The RL agent is trained to learn paths to all the house layout destinations based on the topological map, with the rewards used for the training shown in Equation 4.5. An experiment with three different neural network architecture configurations is chosen to determine the best neural network architectures. Table 4.11 shows the three different architectures considered to determine the ideal hidden layer size and its success rate after training for 2500 episodes. The difference in success rate is marginal; this can be because the agent is trained based on numerical data with fewer features. The success rate is plotted and shown in Figure 4.37. The success rate plot can be considered as the learning curve of the agent. From the plot, it can be concluded that the case2 performs better than other cases.

$$r(s,a) = \begin{cases} 100, & s = g \\ -1, & s = \{x \mid x \not\subset (S \cap G)\} \\ -10, & \text{s is previously visited location} \\ -100, & s = U \not\subset S \end{cases} \tag{4.5}$$

The episodic reward collected during the case-2 training process is shown in Figure 4.38 along with zero and threshold reward lines. The threshold reward is calculated based on the difference between the goal reward and the critical path. A critical path is the biggest path (in terms of node count) among efficient paths to the destinations. Analyzing the reward plot reveals two

Table 4.11: Success rate of different DQL architecture configuration

Test Case	Hidden Layer Size		Success Rate
	Neurons	Layers	
Case-1	24	2	53.3%
Case-2	48	2	55.2%
Case-3	24	4	51.4%

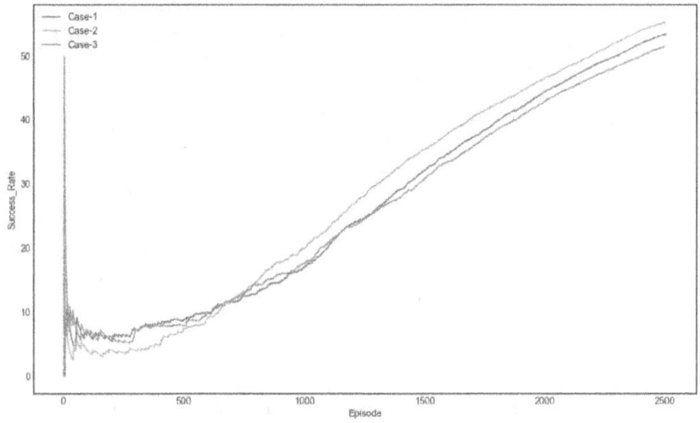

Figure 4.37: Different DQN architectures training success rate

major factors to evaluate the minimum number of training episodes required to train the agent effectively. First, the minimum number of episodes required for training is based on the frequency reduction of the negative rewards about the zero reward line. Second, trainigng efficiency is based on the agent by comparing episodic reward about the threshold reward line. The negative reward frequency reduction and increased frequency of attaining episodic rewards about threshold reward are achieved in 2200 episodes. Hence, 2200 episodes are considered the minimum training episode to train the RL agent in the house layout effectively.

Hospital Layout Training

To verify the scalability of the proposed DQL path planner, it trains the hospital layout, which is another service environment. The DQL architecture is the same as the case-2 type in the house layout. The training hyper-parameters are the same as the house layout. A plot on episodic reward during the training process is shown in Figure 4.39. The total number of training episodes required is over 10,000, so the plot on the left is too small to analyze the information. A scaled version of the same plot with the last 600 episodes of training data is depicted on the right. The RL agent negative reward frequency reduction and frequency of attaining episodic rewards to threshold reward is achieved in 11850 episodes. Hence, the nearest upper-rounded number, 12000,

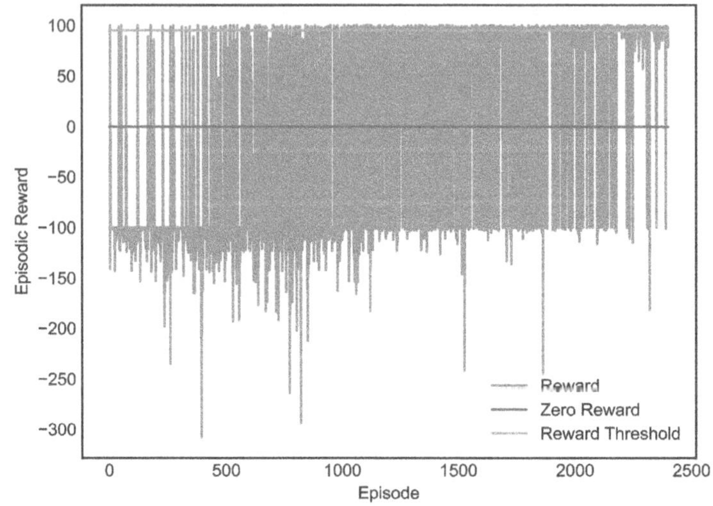

Figure 4.38: House layout case-2 episodic reward

is considered the minimum training episode to train the RL agent in the hospital layout effectively.

In both the service environment, the ϵ exponent decay factor is calculated based on Equation 3.16. For a hospital layout with 12000 episodes of training, the proposed exponent value is 0.9998. Figure 4.40 depicts the training success rate plot for three exponent values. From the plot, it can be concluded that the learning with an exponent value of 0.9998 is efficient compared to other values.

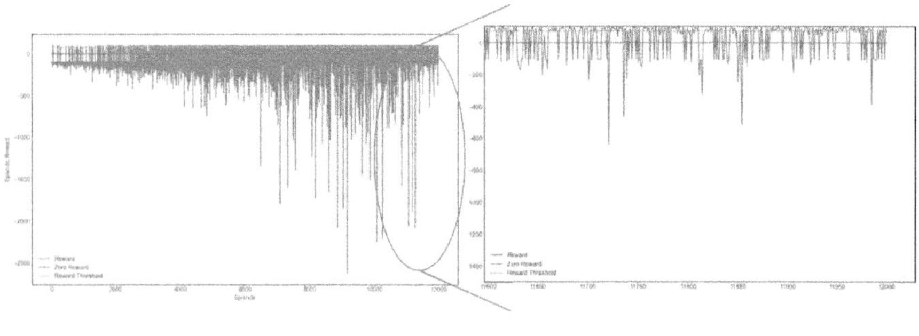

Figure 4.39: Hospital layout episodic reward

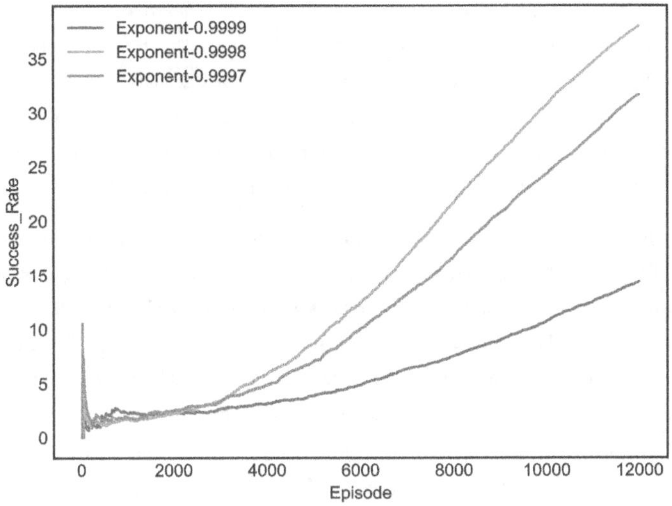

Figure 4.40: Training success rate for different exponents

Table 4.12: Proposed DQL model parameters and generalized parameters

Parameter	House Layout	Hospital Layout	Generalized Parameters
Input Neurons	23	81	$\|S\| + \|G\|$
Output Neurons	4	4	$\|A\|$
Hidden Layer count	2	2	2
Hidden Neurons/layer	48	128	>1.5 x Input Neurons
Batch Size	32	32	Constant
ϵ-Range	1-0.1	1-0.1	Constant
ϵ-Decay Exponent	0.999	0.9998	Eq. (3.16)
Testing Success Rate	100%	98.4%	-

Generalized Parameters

The DQN architecture parameters are chosen based on the topological map properties for successfully training the RL agent in two different service environments. Table 4.12 tabulates all the model parameters, training success rate, and generalized parameters for both environment models. The generalized parameter for all the DQN parameters is derived as the function of topological map properties.

Testing

The proposed DQL path planner is successfully trained in house and hospital environments. To prove that the model is generic and scalable, it is essential to test the model for its ability to plan paths for various destinations given a source. All the destinations in the topological map are considered to be to be sources and destinations for path planning. The model is evaluated based on the testing success rate, calculated as the ratio of the successful and total number of test cases. The testing success rate is more than 98% for both environments.

To further verify the correctness of the generalized parameters mentioned in Table 4.12, the proposed RL agent is tested using a 2D pathfinding maze benchmark map (Sturtevant, 2012). These benchmark maps have been used by researchers worldwide for evaluating path-planning algorithms (Low et al., 2022; Lei et al., 2022; Toma et al., 2021; Zhao et al., 2020). Figure 4.41 depicts the benchmark maze environment with the paths and random destinations. The topological map for the benchmark maze is shown in Figure 4.42 with 18 destinations and 88 waypoints, including home and unknown nodes. There are 108 nodes, 7x and 2x times bigger than the house and hospital layouts. The topological map is too big to fit on a single page and split into three parts. The DQN architecture to train the RL agent for the maze environment is based on the generalized parameters tabulated in Table 4.12. The input layer size n for hospital layout is calculated to be $n = 126(108(\text{states count}) + 18(\text{goal states count}))$, two hidden layers with the size of 190 neurons, and output layer size $m = 4$ are chosen. The training hyper-parameters are the same as in Table 4.10. The RL agent was trained for 56000 episodes with the ϵ-decay factor as 0.999956. The episodic reward plot is shown in Figure 4.43. The episodic reward was increasingly better as the training episodes reached 56000. The testing success rate is 98.3% for the maze layout. This is on par with the hospital and the house layouts. Based on the generalized parameters in Table 4.12, it can be concluded that the proposed DQL system is general and scalable.

Li et al. (2017) proposes a modified Q-learning approach to overcome the drawback of increased convergence time of the traditional Q-learning approach. In this article, the authors have used the benchmark maze environment to compare their proposed approach with traditional Q-learning and other path-planning approaches. The system was trained from a fixed starting point and fixed ending point for over 30 runs, and the averaged comparison metrics were evaluated. The starting point considered is the same as the Home location, and the ending location is D18. This work uses a topological map with 19 destinations to train the RL agent. The final path generated by the RL agent from Home to D18 is shown in Figure 4.44, and it is the same as the modified Q-learning approach (Li et al., 2017).

The house layout DQL agent is also tested in the mobile robot Turtlebot3 using the ROS framework and the Gazebo simulator. Figure 4.45 depicts the two paths, Home to D5 and Home to D4, autonomously generated by the path planner agent.

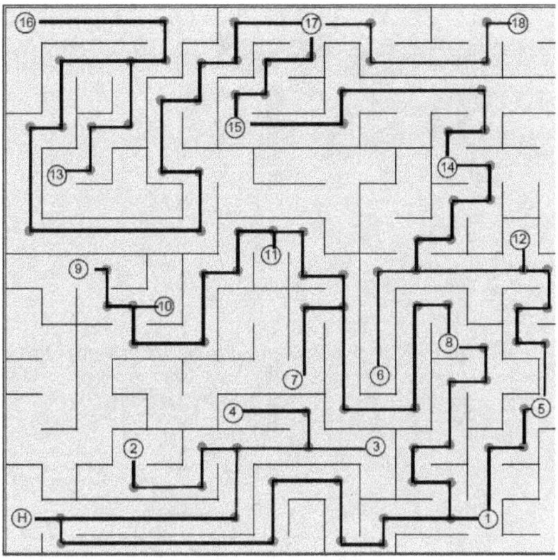

Figure 4.41: 2D pathfinding benchmark maze environment with paths

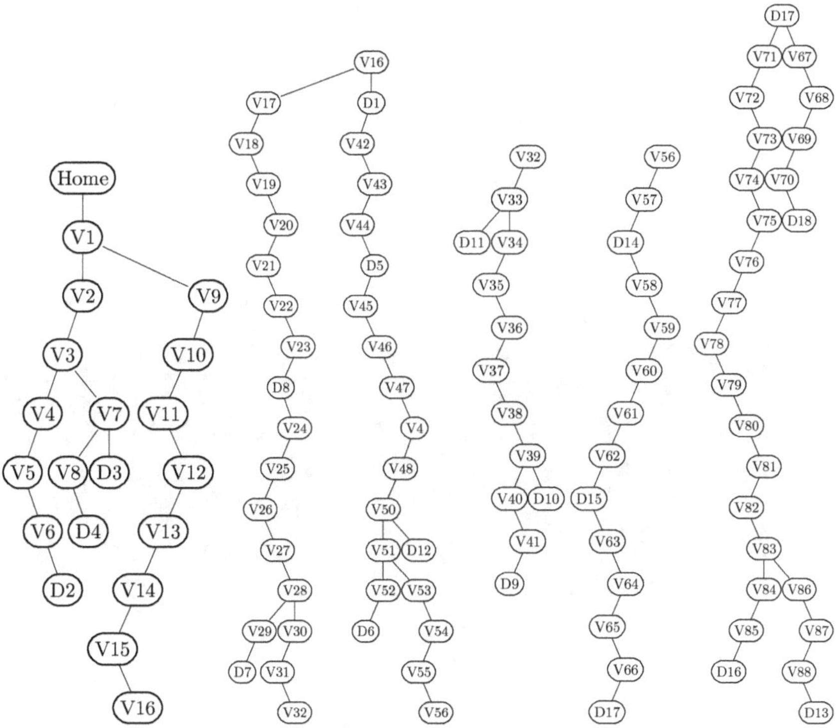

Figure 4.42: Benchmark maze environment path in four parts

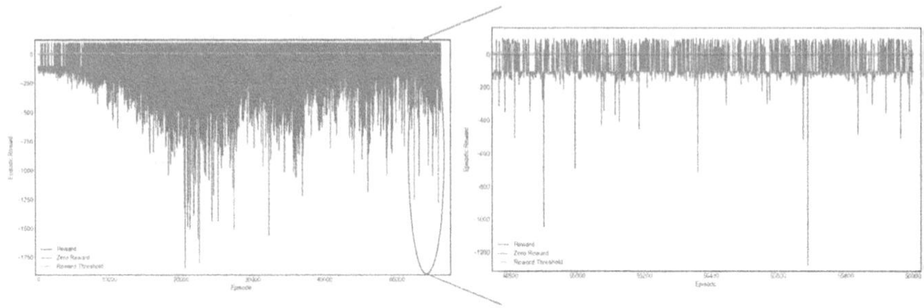

Figure 4.43: Benchmark maze layout episodic reward[Left]

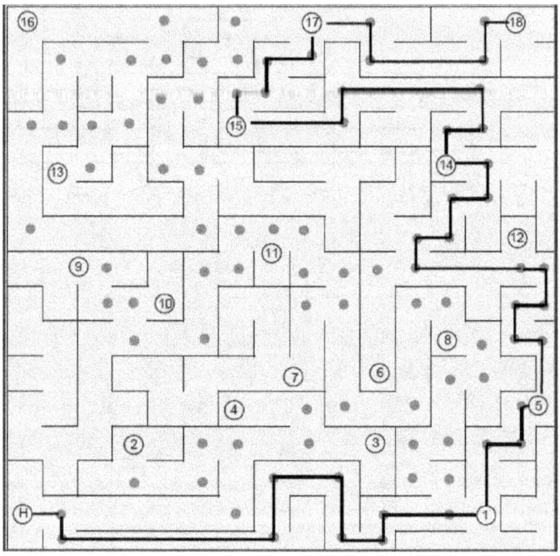

Figure 4.44: Path generated from Home to D18 in benchmark maze environment

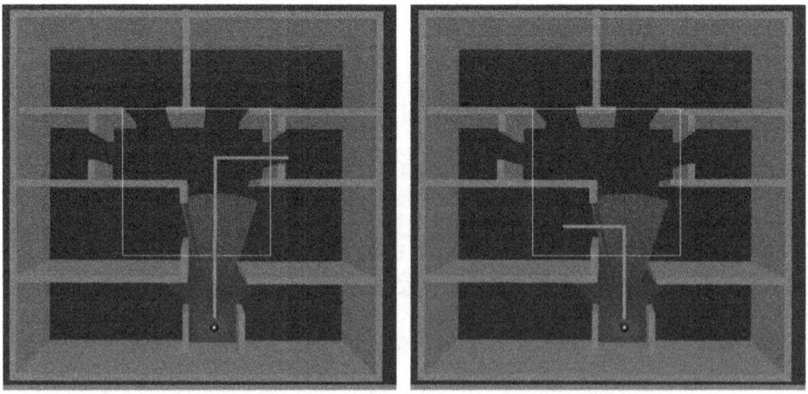

Figure 4.45: Simulator paths(in red) for home to D5 (left) and D4 (right)

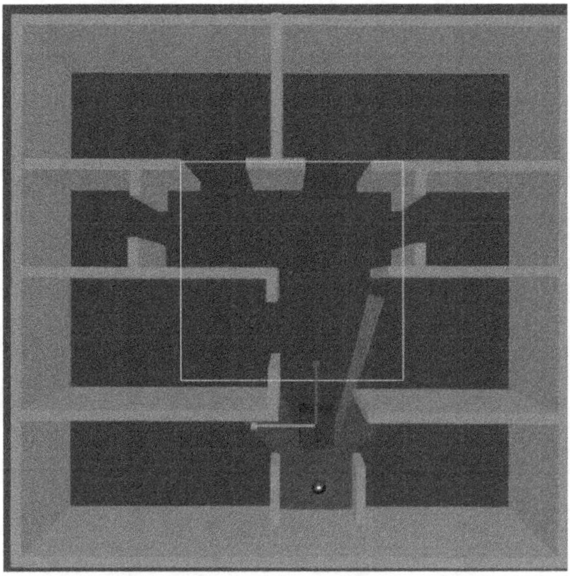

Figure 4.46: Obstacle added between V1 and V2 in house layout

Table 4.13: Testing success rate comparison of TL algorithms

Algorithm	Testing Success Rate
No-beta	7.4%
Constant-beta	41%
Proposed beta-decay	76%

Transfer Learning and Testing

The proposed β-decay transfer learning algorithm is tested by adding an obstacle in the house layout gazebo environment. The obstacle is placed between the V1 and V2 nodes of the house layout. Figure 4.46 and Figure 4.47 depict the house layout with the cuboid obstacle in the Gazebo simulator environment and its topological map. The number of nodes in the topological map is increased, and all the new nodes are highlighted in red with dotted edges for their connections. This changes the environment's configuration, making the pre-trained DQL agent inefficient.

The β-decay transfer learning algorithm transfers knowledge of the source path planner (pre-trained agent) to the target path planner (new agent). The β and ϵ values are calculated as $\beta = 0.78$ and $\epsilon = 0.9984$. The DQL agent is trained using three different TL algorithms, without-β, constant-β, and proposed β-decay. Table 4.13 shows the training success rate for the three TL algorithms after 500 training episodes. The proposed β-decay TL algorithm outperforms the other two.

The DQL agent is trained with and without transfer learning to evaluate the proposed transfer learning approach quantitatively. The metrics to evaluate the efficiency of the TL algorithm are

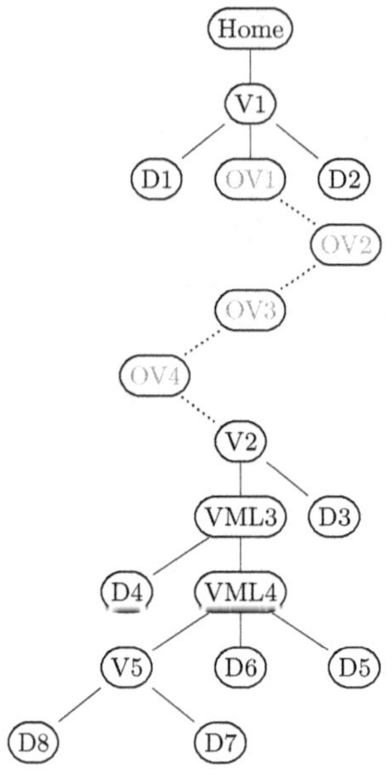

Figure 4.47: Topological map of house layout with obstacle

jumpstart, time-to-threshold, and time-to-converge. Jumpstart is the agent's success rate in the initial few training episodes, and it can be calculated based on the average success rate of the first 100 training episodes. The time-to-threshold metric evaluates the learning performance of the agent by the time it takes to reach the threshold performance. The time-to-converge metric indicates the minimum number of training episodes required to learn the environment for path planning. The first two metrics are evaluated from the training success rate, and the third is from the testing success rate.

Figure 4.48 shows the training success rate plot. The plot shows that the training with transfer learning outperforms the non-transfer training. The threshold value is 42.5%, which is chosen based on the agent training in the old environment. There is only a small incremental change in the new environment compared to the old one. Table 4.14 shows the success rate comparison between transfer and non-transfer learning. The transfer learning approach attains a 100% success rate after 1300 episodes, whereas non-transfer learning attains a 78.5% success rate after 2500 episodes.

Table 4.15 compares all the evaluation metrics between transfer and non-transfer DQL agents with performance improvement factors. In all the metrics, the DQL agent with proposed β-decay transfer learning outperforms the non-transfer agent. The convergence time of the transfer agent is more than two times faster than the non-transfer agent.

The DQL agent with the proposed system is tested in the Gazebo simulator to navigate two paths:

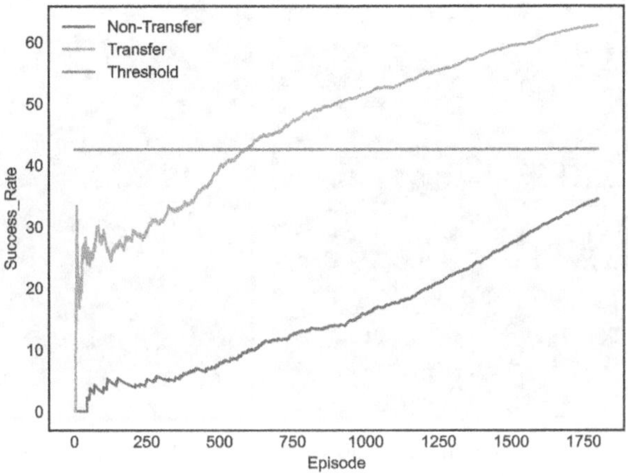

Figure 4.48: Training success rate for transfer vs. non-transfer training

Table 4.14: Testing success rate of transfer and non-transfer training

Episodes	TL	Non-TL
500	26.7%	21.4%
1000	79%	69%
1100	89%	71%
1200	91%	72%
1300	100%	73%
1400	100%	74%
1500	100%	75%
2000	100%	75%
2500	100%	79%

Table 4.15: Transfer learning evaluation metrics comparison

Metric	Units	Non-Transfer	Transfer	Improvement Factor
Jumpstart	Training Success Rate	1%	20%	20x
Time to Threshold	Training Episodes	>1800	≈600	>3x
Time to Converge	Training Episodes	>2500	≈1300	>1.9x

home to D5 and home to D4. Figure 4.49 depicts the paths planned by the path planner by successfully avoiding the obstacles.

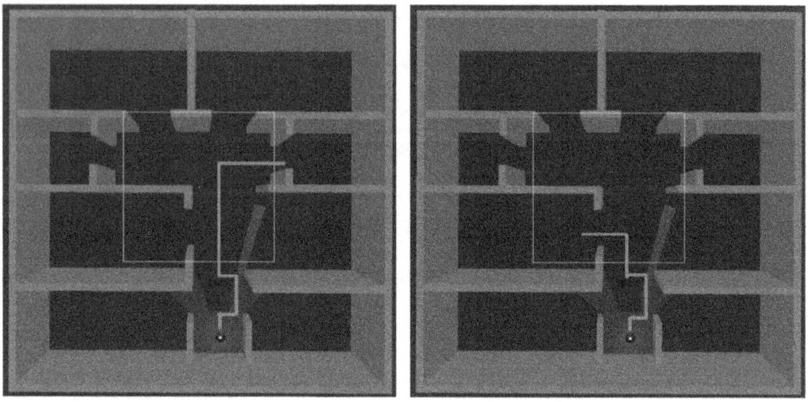

Figure 4.49: β-decay agent simulator paths(in red) for for Home to D5 (left) and D4 (right) in new environment (with obstacle)

A DQL based path planner with novel β-decay transfer learning for lifelong learning ability is proposed and implemented. The DQL path planner is initially pre-trained in a service environment using its topological map. If there are any changes in the environment's configuration, the proposed TL algorithm trains a new DQL agent by efficiently transferring the knowledge of the old agent.

4.4 Conclusion

The proposed navigation framework and its major systems are tested successfully. The framework's scalability is tested by performing the test in three environments that differ in size and number of possible destinations. The dynamic adaptation is tested by changing the environment after the RL agent is pre-trained with the old environment configuration. The test results prove that the RL path planner can adapt to environmental configuration successfully.

Chapter 5

Conclusion and Future Direction

This research focuses on designing and developing a navigation system suitable for service applications. The prime focus is developing a dynamic and self-adapting navigation system for indoor autonomous mobile service robots.

5.1 Conclusion

The proposed navigation framework is divided into three major modules: the Learning from Demonstration (LfD) module, the path mapping and planning module, and the Reinforcement Learning (RL) module. The LfD technique is used to train the robot with at least one path to all the desired destinations. Three LfD techniques are developed incrementally, based on the shortcomings of their predecessors, and are proposed and implemented. The Enhanced Encoder Based Learning (EEBL) technique proved to be the best of all the proposed LfD techniques. A simple user interface is used for user demonstration. On-board sensors are used to identify the path variables while training. This makes the system environment independent. The EEBL algorithm can also correct positional/orientation errors due to hardware anomalies in the robot. The system is tested in two different environments of different sizes. This makes the proposed algorithm portable and scalable.

The path mapping and planning module maps all the logical paths based on the learned path. The proposed Tree-Based Planner (TBP) algorithm maps all the logical paths and builds a path tree by autonomously identifying Virtual Land Mark and reference points. Finally, the path planned from a fixed source and destination by the TBP module is compared with other classical and state-of-the-art path planning techniques. The result shows that the proposed algorithm has fewer visited cells and is on par concerning path length. The proposed state-based obstacle avoidance technique can efficiently navigate around obstacles of any shape and return to the planned path.

The path planned by TBP is static, and the paths will no longer be valid if the environment changes. The proposed Reinforcement Learning (RL) module uses a deep Q-learning-based path planner with novel β-decay transfer learning for lifelong adaptation. The Deep Q-Learning (DQL) agent is trained in two service environments: house and hospital layouts and one benchmark maze environment with a testing success rate of 100%, 98%, and 98.5%, respectively. The hospital and maze environments are almost 3x and 7x larger than the house environment, respectively, concerning the number of nodes and destinations. This proves that the proposed path planner is scalable. The generalized architecture parameters of the DQL agent are derived from the environment topological map. The path planner was successfully tested in the ROS framework with the Gazebo simulator using the Turtlebot3 mobile robot.

Learning lifelong is tested by adding an obstacle in the home environment. The proposed Transfer Learning (TL) algorithm uses the Experience vs. Exploration vs. Exploitation (EEE) logic based on the β-decay factor. The choice of the initial value of β and decay factor is evaluated by comparing it with non-β TL and constant β TL. The β TL algorithm performed better than the others. The efficiency of the TL algorithm is evaluated based on the metrics jumpstart, time-to-threshold, and time-to-converge compared to the non-TL method. The proposed TL method's jumpstart is improved by a speedup factor of 20, time-to-threshold was achieved three times faster, and convergence time was reduced by half.

Hence, based on the experimental results, the proposed learning-based navigation system is environment-independent, portable, dynamic, and best suited for indoor service environments. The proposed navigation framework is designed to be easily implemented in real robots. The major characteristics of the system that are proved experimentally in this thesis are the key factors for real-time implementation. Environment independence makes the real robot rely only on onboard sensors like the Proximity sensor, pose sensor, and wheel encoder. It does not need any external sensor from the environment. The proposed framework allows the robot to be portable to any environment easily. After porting, the robot has to be trained in the environment, after which the robot can operate autonomously. The framework allows the robot to work in a dynamic, changing environment for a longer time, thanks to the lifelong learning ability of the proposed framework.

5.2 Future Direction

The usage of service robots in our day-to-day lives has increased significantly in the past decade. This growth trend will continue, and service robots like mobile phones and computers will soon be part of our lives. In this work, the LfD module uses proximity sensors and wheel encoders to learn the path from human demonstrations. Image data can identify landmarks and reference points like doorways and corridors. This approach's inherent challenge is identifying the environment's features using the camera mounted on the robot.

Here, RL module enables the system to change the path planning strategy based on environmental changes dynamically. The TL technique plays a key role in updating the path planning efficiently with less training time and using the old learned information as much as possible. Methods to improve the efficiency of the TL algorithm can be explored. The β factor is a stochastic factor initialized based on the total number of similar states in the environment identified based on the state names. Identifying the similarity based on state variables can increase the efficiency of the TL algorithm.

The overall navigation system's efficiency relies on the initial paths learned based on human demonstrations. One key assumption is that the demonstrations are the shortest path possible in the environment. However, this assumption is realistic and very obvious. As a research problem, one can always explore the possibility of identifying the shortest path based on inefficient human demonstrations. This is an open problem that can lead to interesting and useful findings. Another open problem is to use RL based imitation learning instead of LfD based learning. This can be another approach to obtain dynamic path planning efficiently. The proposed navigation framework is designed to be implemented in real-time. In this aspect, a detailed study of the RL agent's performance evaluation and hardware requirement to implement such a system in real life can be one of the future directions.

The proposed framework can be updated to incorporate socially aware navigation functionality. The system is proposed to work in a service environment where robots operate alongside humans. Incorporating human motion trajectory and other social aspects for service robot path planning is important. Lifelong learning combined with socially aware navigation makes the navigation framework more usable and robust for service environments.

www.ingramcontent.com/pod-product-compliance
Lightning Source LLC
LaVergne TN
LVHW010737061126
851857LV00014B/1516
9781962116862